James Alfred Wanklyn, Ernest Theophron Chapman

Water-Analysis

A Practical Treatise on the Examination of Portable Water

James Alfred Wanklyn, Ernest Theophron Chapman

Water-Analysis
A Practical Treatise on the Examination of Portable Water

ISBN/EAN: 9783744650076

Printed in Europe, USA, Canada, Australia, Japan

Cover: Foto ©berggeist007 / pixelio.de

More available books at **www.hansebooks.com**

WATER ANALYSIS:

A PRACTICAL TREATISE

ON THE

EXAMINATION OF POTABLE WATER.

BY

J. ALFRED WANKLYN, M.R.C.S.,

PROFESSOR OF CHEMISTRY IN THE LONDON INSTITUTION,

AND

ERNEST THEOPHRON CHAPMAN.

LONDON:

TRÜBNER & CO., 60 PATERNOSTER ROW.

1868.

TO THE

PROPRIETORS OF THE LONDON INSTITUTION,

IN WHOSE LABORATORY THE WORK DESCRIBED

IN THE

FOLLOWING PAGES WAS DONE,

This Book

IS

RESPECTFULLY DEDICATED BY

THE AUTHORS.

PREFACE.

THIS book is designed to give an account of that kind of Water Analysis which is proper for the examination of water intended for domestic use. In giving processes of analysis, we have been careful to prescribe only such operations as are practicable. Especially we have kept in view the necessity of economy in time. The complete examination of a sample of water, as set forth in the following pages—viz., the determination of solid residue, hardness, chlorine, nitrogen as nitrates and nitrites, ammonia, and organic matter—may be all done in the space of six hours, whilst a partial examination takes a very short time. In thus economising time, we do not think that we have made any sacrifice of accuracy.

The process for determining nitrogenous organic matter, termed by us the *ammonia-method*, and proposed by Mr Miles H. Smith, and ourselves, at a meeting of the Chemical Society last year, occupies a prominent place in this work. A section of Chapter V. is devoted to a discussion of the interpretation to be given to the data furnished by that process. The process by Frankland and Armstrong is not given in the text. We believe that the great length of time and great skill requisite to execute it would render it quite impracticable, even if it were otherwise desirable. Moreover, we do not set any value on its indications. Our objections to it are given in the Appendix.

A prominent feature of this work is its copiousness in examples, more especially in examples of the new determination of nitrogenous matter. In fact, in this time of Governmental water-commissions of different kinds, we have issued a kind of private water-commission of our own, availing ourselves of the resources of the ammonia-method report on the state of the London, Manchester, Edinburgh, and Glasgow water-supply during last summer.

We believe that this is the first book which has been published on Water Analysis. Up to the present time, the subject has been treated of in a special section in works devoted to *chemical analysis in general ;* as, for instance, in the admirable analytical treatise by Fresenius, and in Dr Noad's excellent manual, from which we quote *verbatim* an account of Clark's method of determining the hardness of water. The subject has also been handled in the "Handwörterbuch," and in Watts' "Dictionary of Chemistry," wherein will be found a very admirable article on Water, written by Dr Paul. "Discourses" on the Analysis of Potable Water have also been given to the Chemical Society by Dr W. A. Miller, and by Dr Frankland and Mr Armstrong, and are to be found in the Journal of that Society.

We have freely availed ourselves of all material within our reach. In particular, to Dr Miller's "Discourse" just referred to, we are indebted for an account of the preparation of the Nessler test. This test, which we believe to be the most delicate known to chemists, was first rendered quantitative by the late Mr Hadow and Dr Miller. The account of the test, which appeared in the "Laboratory" last year, was taken from Dr Miller's "Discourse." The account given in this work is essentially Dr Miller's, being slightly modified, however, in accordance with the dictates of our own rather large experience in the practical working of the test.

To Dr Frankland we are indebted for the account of the determination of nitric acid by the modification of Walter Crum's method. This we have quoted from the Journal of the Chemical Society, giving it in addition to the determination of nitric acid by the modification of Schultze's aluminium process, which we prefer to recommend. We agree with Dr Frankland in the light estimation in which he holds the "loss on ignition." We go with him, also, a long way in his condemnation of the "permanganate test" as a trustworthy measure of the organic impurity of water.

In other respects relating to the water analysis we are for the most part at variance with him.

On the subject of the invalidity of the determination of organic matter in water by Frankland and Armstrong's combustion process, we believe that chemists in general agree with us. A considerable part of this book is new in every sense of the word. The treatment to which we have subjected Clark's table is novel. The fact of the extreme constancy (organically considered) of the good water supply all over the country is here announced for the first time. The great advantage of filtration and of the application of Clark's softening process to the purification of water is brought out in this work. The fact that the carbonate of lime carries down albumen in preference to such substances as urea is new. The estimations of nitric acid and ammonia without distillation are new. Other things that are new will be found by our readers.

LABORATORY OF THE LONDON INSTITUTION,
May 1868.

CONTENTS.

CONTENTS.

WATER ANALYSIS.

INTRODUCTION.

(1.) *The Collection of the Sample of Water.*—The kind of glass bottle known as a " Winchester Quart" (being a stoppered bottle holding 6 lbs. or about 3 litres) will answer very well for the purpose of holding the sample of water. A half-Winchester bottle, or two stoppered pint-bottles, will, in case of necessity, admit of an examination being made.

(2.) The bottle should be cleaned out with strong sulphuric acid, and then be washed with ordinarily good water, until the washings are no longer acid. Before being charged with the sample of water, the bottle should be rinsed out with some of the same kind of water as that which is intended to be collected for examination. After being filled (almost, but not quite full up to the neck) with the sample of water, the bottle should be stoppered and kept in a cool and dark place until it is examined. It is desirable to make the examination for organic

A

matter as soon as possible; within forty-eight hours after collection if possible.

(3.) In taking the water-supply of a town, it is well to take the water direct from the street mains. In London there are water-jets at the cab-stands which are particularly well adapted for the purpose of getting samples of the water of the different companies.

It will be understood that, before taking a sample, a little water should be allowed to flow so as to clear the pipe. In taking a river—or pond—water, the bottle should be immersed in the water, the mouth of the bottle being some distance under the surface, so as to avoid getting any scum into the bottle. In taking river-water, select the middle of the stream, and avoid the outlets of sewers and feeders: note whether there has been heavy rain or long drought.

(4.) The stopper of the bottle should be tied over with a piece of clean linen or calico, and the string sealed. The employment of linseed meal round the stopper is to be avoided.

When the analysis is made, it should be noted whether the water has been filtered or not; also the physical characters should be noted, whether it is bright and clear, or turbid—whether it is coloured or not—whether it has any smell. Its reaction towards litmus may also be examined:

To 100 cubic centimeters (about 3 oz.) of the water, 1½ cubic centimeters of Nessler reagent may be added. The occurrence of a brown colour, or brown precipitate, is in itself a sufficient reason for condemning a water.

(5.) The chief chemical data given by such an examination of water as it is desirable to make for sanitary purposes, are the following :—

    I. Total solid residue.

    II. Hardness, temporary and permanent.

    III. Chlorine.

    IV. Nitrogen, existing as nitrates and nitrites.

    V. Ammonia, and organic matter.

    VI. Metals.

We shall accordingly, in the following chapters, treat each of these data in succession.

CHAPTER I.

TOTAL SOLID RESIDUE.

(6.) THE determination of the amount of total solids in water is a very simple operation. A known quantity of water is evaporated to dryness, and the residue dried and weighed. It is usual to take half a litre of water or more, one or two litres being sometimes recommended. The evaporation ought to be begun and finished in one and the same platinum vessel. The practice of beginning the evaporation in a porcelain vessel, and afterwards transferring the concentrated water to a small platinum dish, is to be condemned.

(7.) We prefer to operate on much smaller quantities of water than are usually taken, and thereby effect a great saving of time without any sacrifice of accuracy. A platinum dish is used, capable of holding 100 c. c. without being full, up to much within one-fourth of an inch of the edge, and provided with a lid of platinum foil or a mica plate. The operation is conducted as follows :—

The platinum dish is cleaned, wetted, wiped, heated to about 130° C., covered with its lid, and then cooled by being placed for a minute on the clean surface of a massive piece of iron.* Having been thus got into exactly the same state as it will be in after the evaporation of the water, the platinum dish is weighed. Either 70 c. c. or 100 c. c. of the sample of water is next put into it, and evaporated in the water-bath or steam-bath. A convenient form of steam-bath, used in the laboratory of the London Institution, is made out of a common two-gallon tin can, such as is sold for containing methylated spirit, varnish, &c. The neck of the can is fitted with a perforated cork. Through the perforation in the cork passes the neck of a wide-necked glass funnel. This funnel may be either plain or ribbed, and is designed to support the platinum-dish, which is made to rest in the wide mouth of the funnel. The escape of the steam is provided for by the insertion of a little roll of filter paper between the dish and the mouth of the funnel, when a plain instead of a ribbed one is employed. The advantages of this form of steam-bath are extreme cleanliness, and also that it may be used for a long time without being replenished.

(8.) When the water-residue in the platinum-dish appears dry, the bottom of the dish is wiped and the dish transferred to the air bath, which is

---

* A flat smoothing iron, with the handle knocked off, will answer very well.

heated to about 130° C. Sometimes the water-residue shows a tendency to spirt on being heated to 130° C. To get over this difficulty, it is advisable to begin the drying with the *lid on*. The lid should, however, not be kept on all the time that the drying lasts. If the temperature of the air-bath is about 130° C. when the dish containing the residue is put into it, a very short sojourn in the air-bath will be sufficient to effect complete drying. After being dried at 130° C., the platinum-dish, containing water-residue and covered with its lid, is cooled by a short contact with the mass of cold iron before mentioned. It is then weighed, and the difference between this weighing and the one before mentioned, gives the amount of total solid residue contained by the water.

If 70 c. c. has been taken, one milligrm. of water-residue will be equivalent to one grain per gallon.

If 100 c. c. has been taken, $\frac{1}{10}$ milligrm. will be equivalent to one part of residue in one million parts of water.

(9.) By operating in the manner described, a determination of the total solid residue in water may be made in about an hour and a quarter; and if the experimenter be in possession of a fair balance and fairly accurate weights, he will easily obtain results not varying by more than half a grain per gallon.

We give examples. 100 c. c. of a sample of water left ·1027 grm. of residue. In a second experiment on the same kind of water, 100 c. c. left ·1030 grm. of residue. From which is calculated,

(1.) Total solids per gallon equal 71·89 grains.
(2.) ,, ,, ,, 72·10 ,,

Another sample of water yielded the following results. 100 c. c. gave ·0352 grm. 100 c. c. gave ·0359 grm. of residue. From which is calculated,

(1.) Total solids per gallon equal 24·64 grains.
(2.) ,, ,, ,, 25·13 ,,

Even in the case of water which, like some lake-waters, leaves a very small residue, we do not think that the taking of larger quantities of water conduces to accuracy.

(10.) The evaporation of large quantities of water involves error, from dust and atmospheric impurities getting into the water; it also involves a certain amount of destruction of the organic matter present in water; and, in short, the advantage gained by having a larger water-residue to weigh appears to be counterbalanced—and, we think, in many instances, more than counterbalanced—by the error arising from the prolongation of the process of evaporation.

We should not, except under very special circumstances, take more than 70 c. c. or 100 c. c. of any specimen of water for the determination of the total solid residue.

(11.) The addition of a small and accurately measured quantity of carbonate of soda to the water to be examined has been recommended by some chemists. We do not consider that any advantage is to be gained by it, so far as the mere determination of the amount of water residue is concerned. The reason why it was recommended had reference to the determination of organic matter by ignition, and will be discussed in Chapter V.'

(12.) In order to give an idea of the amount of solid residue which actually occurs in natural water, we subjoin a few examples taken from the large table at the end of Watts' Dictionary:—

	Grains per gall.	Parts per million. (Milligramme per litre.)
Loch Katrine, Glasgow	2·30	32·8
Bala Lake water (Frankland and Odling)	1·95	27·9
Bala Lake water (Smith)	3·18	45·4
Manchester water	4·76	68·0
Thames water supplied to London	21·66	309·4
Lake of Geneva	10·64	152·0
Rhine at Basle	11·86	169·4
Spree at Berlin	7·98	114·0
Atlantic Ocean	2688·00	38400·0
Distilled water (Fresenius)	0.12	1.7

It will occur to our readers, that, excepting sea water, natural water is a substance of high purity.

Average London water from the Thames, for instance, contains about 0·03 per cent. of impurity. A chemist considers himself fortunate when he meets with a chemical compound so little contaminated with foreign matter as to contain 99·97 per cent. of that compound.

(13.) We are at present very much in the dark with regard to the sanitary aspect of the amount of solids existing in water. Whether a water having an exceptionally small amount of solid contents is specially salubrious, remains to be ascertained. A very large quantity of fixed matter is certainly unwholesome. To take an extreme case, sea-water is absolutely non-potable. But whether the 21 grains per gallon present in London water, and whether double that amount, would do the smallest damage to the health of persons who should drink such water, is an open question.

CHAPTER II.

HARDNESS.

(14.) ONE of the most important domestic uses of water is to wash with, and one of the most striking differences between different kinds of natural water is the behaviour of it towards soap. Some water destroys much soap before a lather is formed; such water is said to be hard. Other kinds of water admit the formation of a lather almost instantly, and are said to be soft.

(15.) The destruction of soap is due to the formation of insoluble salts, by reaction between the lime and magnesia of the water and the soap. Not until the lime and magnesia salts present in the water have exhausted themselves upon the soap, (forming insoluble lime or magnesia salts of the fatty acids of soap,) will there be formation of lather. In measuring the hardness of water, there are two obvious ways of constructing the scale of hardness. We may measure the amounts of soap destroyed by a volume of different kinds of water; or, bearing in mind the cause of hardness in a water, we might measure the amount of fixed matter in a volume of different kinds of water. On the one plan, the degree of hardness is the quantity of soap destroyed —so much soap destroyed, so many degrees of hardness. On the other plan, the degree of hardness is the quantity of carbonate of lime (or its equivalent of other salts) in the water. Both methods of registering hardness are intelligible, but the former is direct; and regard being had to the fact that it is the actual consumption of soap that we want practically to know, it is better to regard "degrees of hardness" as quantities of soap destroyed, and not as quantities of carbonate of lime which do damage by destroying soap.

(16.) We transcribe the following account of the soap test from Dr H. M. Noad's valuable little book on *Quantitative Analysis*, p. 576 :—

"(*a*) We are indebted to Dr Clark for a very simple method of determining the *degree* of hardness of a water. It consists in ascertaining the quantity of a standard solution of soap in spirit required to produce a permanent lather with a given quantity of water under examination; the result being expressed in degrees of hardness, each of which corresponds to 1 grain of carbonate of lime in a gallon (=70,000 grains of distilled water) of the water. From the specification of his patent (enrolled 8th September 1841) we gather the following particulars :—

(*b*) *Preparation of the Soap Test.*—Sixteen grains of pure Iceland spar (carbonate of lime) are dissolved (taking care to avoid loss) in pure hydrochloric acid, the solution is evapo-

rated to dryness in an air-bath, the residue is again redis-
solved in water, and again evaporated; and these operations
are repeated until the solution gives to test-paper neither
an acid nor an alkaline reaction. The solution is made up
by additional distilled water to the bulk of precisely one
gallon. It is then called the "standard solution of 16 degrees of hardness." Good London curd soap is dissolved in
proof spirit, in the proportion of one ounce of avoirdupois
for every gallon of spirit, and the solution is filtered into a
well-stoppered phial, capable of holding 2000 grains of distilled water; 100 test measures, each measure equal to 16
water-grain measures of the standard solution of 16 degrees
of hardness, are introduced. Into the water in this phial
the soap solution is gradually poured from a graduated
burette; the mixture being well shaken after each addition
of the solution of soap, until a lather is formed of sufficient
consistence to remain for five minutes all over the surface of
the water, when the phial is placed on its side. The number of measures of soap solution is noticed, and the strength
of the solution is altered, if necessary, by a further addition
of either soap or spirit, until exactly 32 measures of the
liquid are required for 100 measures of the water of 16 degrees of hardness. The experiment is made a second and a
third time, in order to leave no doubt as to the strength of
the soap solution, and then a large quantity of the test may
be prepared; for which purpose Dr Clark recommends to
scrape off the soap into shavings, by a straight sharp edge of
glass, and to dissolve it by heat in part of the proof spirit,
mixing the solution thus formed with the rest of the proof
spirit.

(c) *Process for ascertaining the Hardness of Water.*—Previous to applying the soap test, it is necessary to expel from
the water the excess of carbonic acid—that is, the excess
over and above what is necessary to form alkaline or earthy
bicarbonates, this excess having the property of slowly
decomposing a lather once formed. For this purpose, before
measuring out the water for trial, it should be shaken
briskly in a stoppered glass bottle half filled with it, sucking

out the air from the bottle at intervals by means of a glass tube, so as to change the atmosphere in the bottle; 100 measures of the water are then introduced into the stoppered phial, and treated with the soap test, the carbonic acid eliminated being sucked out from time to time from the upper part of the bottle. The hardness of the water is then inferred directly from the number of measures of soap solution employed, by reference to the subjoined table. In trials of waters above 16 degrees of hardness, 100 measures of distilled water should be added, and 60 measures of the soap test dropped into the mixture, provided a lather is not formed previously. If, at 60 test measures of soap test, or at any number of such measures between 32° and 60°, the proper lather be produced, then a final trial may be made in the following manner:—100 test measures of the water under trial are mixed with 100 measures of distilled water, well agitated, and the carbonic acid sucked out; 'o this mixture soap test is added until the lather is produced, the number of test measures required is divided by 2, and the double of such degree will be the hardness of the water. For example, suppose half the soap test that has been required correspond to $10\frac{5}{10}$ degrees of hardness, then the hardness of the water under trial will be 21. Suppose, however, that 60 measures of the soap test have failed to produce a lather, then another 100 measures of distilled water are added, and the preliminary trial made, until 90 test measures of soap solution have been added. Should a lather now be produced, a final trial is made by adding to 100 test measures of the water to be tried, 200 test measures of distilled water, and the quantity of soap test required is divided by 3; and the degree of hardness corresponding with the third part being ascertained by comparison with the standard solutions, this degree multiplied by 3 will be the hardness of the water. Thus, suppose 84·5 measures of soap solution were required $\frac{84\cdot5}{3} = 28\cdot5$, and on referring to the table this number is found to correspond to 14°, which, multiplied by 3, gives 42° for the actual hardness of the water."

(*d*) Table of Soap-test Measures corresponding to 100 test measures of each standard solution :—

Degree of hardness.	Soap-test measures.	Differences as for the next degree of hardness.
0	1·4	1·8
1	3·2	2·2
2	5·4	2·2
3	7·6	2·0
4	9·6	2·0
5	11·6	2·0
6	13·6	2·0
7	15·6	1·9
8	17·5	1·9
9	19·4	1·9
10	21·3	1·8
11	23·1	1·8 .
12	24·9	1·8
13	26·7	1·8
14	28·5	1·8
15	30·3	1·7
16	32·0	

(17.) From this it appears that Dr Clark has adopted the indirect and not the direct way of registering hardness (or soap-destroying power.) According to him, the hardness of a water is measured by the quantity of carbonate of lime (or its equivalent in other salts) which the water contains. and not by the quantity of soap which a gallon of it will destroy.

(18.) Obviously, the indirect method of measuring hardness is only valid, in so far as it is parallel with the direct way.   As will be seen from the table for the reduction of soap-measures to their equivalent in carbonate of lime, there is a certain want of parallelism between the direct and indirect measurement of soap-destroying power. In order to bring this fact out distinctly, we repeat the table in an altered form. We have divided the number of soap measures by 2, and thereby made the last term of the table alike for both carbonate of lime and corresponding soap-measures.

Grains of Carbonate of Lime per Gallon.	Soap-test Measures.
0	0·7
1	1·6
2	2·7
3	3·8
4	4·8
5	5·8
6	6·8
7	7.8
8	8·75
9	9·7
10	10·65
11	11·55
12	12·45
13	13·35
14	14·25
15	15·15
16	16·00

It will be observed that the zero of carbonate of lime is opposite to 0·7 measure of soap test. What is the explanation of this? Obviously the gallon of pure water requires a certain amount of soap, and not infinitesimally little soap in order to yield a permanent lather. Dr Clark's degrees of hardness are grains per gallon of carbonate of lime, or *its equivalent in soap-consuming power*.   In the instance of pure water, the thing equivalent to carbonate of lime in soap-consuming power is the gallon of water.

According to the table, one gallon of pure water destroys as much soap in producing a lather, as 0·8 grain of carbonate of lime would destroy if it were added to a gallon of water, which already contained, say, 10 grains of carbonate of lime, and a small quantity of soap test in excess over that required to neutralise the 10 grains of carbonate of lime.

Making the gallon of pure water count as the equivalent of 0·8 grain of carbonate of lime, we may rectify the table as follows :—

Grains of Carbonate of Lime (or matter equivalent to Carbonate of Lime) in a Gallon of Water.	Measures of Soap Test.
0·8	0·73
1·8	1·68
2·8	2·83
3·8	3·99
4·8	5·04
5·8	6·09
6·8	7·14
7·8	8·19
8·8	9·19
9·8	10·18
10·8	11·18
11·8	12·13
12·8	13·07
13·8	14·02
14·8	14·96
15·8	15·91
16·8	16·80

This table is derived from the foregoing by adding 0·8 grains to the carbonate of lime in every case, so as to correct for the soap-destroying power of the gallon of water; and every reading of soap-test is multiplied by 1·05, so as to make the 16 measures into 16·80, and to increase the others in the same ratio.

(19.) This treatment of the table, (and the logic of it will be evident on examination,) exhibits the destruction of soap as very fairly parallel with the amount of carbonate of lime. We regard the departure from the parallel as within the limit of experimental error. From experimental data of our own we are inclined to rate the soap-destroying power of a gallon of pure water at equal to 1·0 grain, rather than 0·8 grain of carbonate of lime, and that will bring the numbers into closer approximation. We exhibit it. The factor used is 1·063.

Grains of Ca_2OCO_2, or its equivalent in a Gallon of Water.	Soap Measures corresponding to the Ca_2OCO_2.
1	0·74
2	1·70
3	2·87
4	4·04
5	5·10
6	6·17
7	7·23
8	8·29
9	9·30
10	10·31
11	11·32

Grains of Ca OCO_2, or its equivalent in a Gallon of Water.				Soap Measures corresponding to the Ca_2OCO_2.
12	.	.	.	. 12·28
13	.	.	.	. 13·24
14	.	.	.	. 14·20
15	.	.	.	. 15·18
16	.	.	.	. 16·10
17	.	.	.	. 17·00

(20.) This table speaks for itself, and shows that the want of strict parallelism between the carbonate of lime and the soap measures is due to experimental error solely. We may therefore get rid of the correction for supposed want of correspondence between the indications of hardness, as given by the carbonate of lime, and as given by the quantity of soap consumed.

In place of making 16 (*or* 32) *measures of soap test to correspond to* 16 *grains of* Ca_2OCO_2 *in one gallon of water, make the measure* 17 (*or* 34) *of soap test, and the correction for want of correspondence will vanish.*

(21.) We prefer to make the standard solutions in a way somewhat different from that which was recommended by Dr Clark, as given in the passage above quoted.

We find it very convenient to keep a standard chloride of calcium solution, (strength equivalent to 1 milligramme of carbonate of lime in a cub. cent. of solution.)

(22.) This we prepare as follows :—Good chloride

of calcium, which is prepared by dissolving a little marble in pure hydrochloric acid, is fused in a platinum crucible, which has been previously cleaned, ignited, and weighed. From 0·1 to 0·2 grm. of the chloride is taken. After heating the chloride of calcium so as to fuse it, the lid is put on to the crucible, and the whole cooled and weighed. The difference between the weight of the empty crucible and the weight of the crucible containing the chloride of calcium, which has been fused, will be the weight of the chloride of calcium employed. Dissolve it in a little water, and, calculating from the known weight, make up with water so as to get a solution of which one litre contains 1·110 grmme. of chloride of calcium. Each cub. cent. of this solution will therefore contain CaCl, equivalent to 1 milligramme of Ca_2OCO_2. This standard solution of chloride of calcium may, of course, if thought necessary, be verified with standard solution of nitrate of silver, (see Chlorine.)

The basis of the system of standards is thus a standard solution of chloride of calcium : strength 1 c. c. = 1 milligramme of carbonate of lime.

(23.) One of us has made the observation that in testing soap solution against hard water it is of no consequence whether the soap solution be added to the hard water and the incipience of the lather be observed, or whether, conversely, the hard water be added to the soap solution and the vanishing of the lather be observed. The vanishing point is just as sharp and quite as easy to recognise as the point at which lather begins to form.

Such being the case, there are various ways of working the soap test.

We next describe the way of making the *standard soap test.*

(24.) Take two parts of lead plaister and one of carbonate of potash, pound them well together, and exhaust repeatedly with alcohol of about 90 per cent., using altogether about 30 times as much alcohol as lead plaister. It will be found convenient to pound only small quantities of lead plaister and carbonate of potash at one operation. The soap solution obtained, as just described, is a solution of potash-soap, carbonate of lead being the complementary product. Before being used the soap solution should be allowed to stand for some time and then filtered. Before being standardised, it should be diluted with its own volume of water, and will then be still far too strong for use. The advantage of solution of potash-soap over soda-soap is that it does not deposit continually, as soda-soap is apt to do.

In order to standardise the soap solution, and so make the *standard soap test,* so that 1 *cub. cent. is* equivalent to 1 *milligramme of carbonate of lime,* proceed as follows :—

(25.) Measure accurately say 10 c. c. of the unstandardised soap solution, put it into a bottle with 70 c. c. of pure water, and then add the standard CaCl solution (strength 1 c. c. = 1 milligramme Ca_2OCO_3) until frothing stops, care being taken to shake up properly. From the result of the trial calculate how much dilution of the soap solution is requisite in order to make 17 c. c. of the soap test consume 16 c. c. of the CaCl solution; dilute the unstandardised soap solution accordingly with alcohol of 40 per cent., and verify the standard soap test after it has been made up.

17 c. c. of standard soap test should accurately neutralise 16 c. c. of standard CaCl solution in the presence of 70 c. c. of pure water.

Each c. c. of the standard soap solution will then be equivalent to 1 milligramme of carbonate of lime.

The use of this standard for ascertaining the degree of hardness of water will be plain from what has already been said. We give explicit directions as follows :—

(26.) Take 70 c. c. of the water and put it into a proper bottle. Add the standard soap test until lather is formed, and note down the number of c. c. of standard soap test required. Each c. c. of standard soap consumed indicates 1 grain of Ca_2OCO_2 (or its equivalent) in one gallon of the water.

If the result be desired, milligrammes of Ca_2OCO_2 in a litre of water, take 100 c. c. of the water, and each c. c. of standard soap will be equal to 10 milligrammes of Ca_2OCO_2 per litre.

Should 70 c. c. of the water require more than 17 c. c. of standard soap test, it will be requisite to dilute the water with its own volume of distilled water, as has been already explained.

A very neat way of taking hardness is as follows :— Take a known volume of unstandardised soap solution and dilute it with 70 c. c. of pure water, and then note the number of c. c. of standard CaCl solution (1 c. c. = 1 milligramme Ca_2OCO_2) required to *stop* lathering. (The number of c. c. of CaCl solution must approximate to 16.) The number of c. c. of CaCl solution + 1 = the number of milligrammes of *potential* Ca_2OCO_2 required by the volume of soap. (The "1" is for the 70 c. c. of water.) Then take the same volume of soap solution, and add it to 70 c. c. of the water to be examined. *If a lather forms*, note how many c. c. of standard CaCl solution are required to *stop lathering.* The difference between the number of milligrammes of Ca_2OCO_2 required to stop lathering, and the potential Ca_2OCO_2 before named, is the hardness of the water in grains of Ca_2OCO_2 per gallon. *If a lather does not* form on the addition of the 70 c. c. of water, add another volume of the soap test, and 70 c. c. of distilled water, (which will count as one milligramme of carbonate of lime.)

(27.) An example will make this plain.

Took one measure of soap solution, added 70 c. c. of distilled water, and noted that it required 14 c. c. of standard CaCl solution to stop lathering. Therefore potential Ca_2OCO_2 required by the measure of soap solution = 15 milligrammes.

Took one measure of soap solution. It did not lather with 70 c. c. of the sample of water. Added a second measure of soap solution, and 70 c. c. of distilled water; it gave a lather. Added standard CaCl solution; 5 c. c. stopped formation of lather.

$$
\begin{array}{llr}
\text{Total potential } Ca_2OCO_2, & = & 30 \\
(1 + 5) & = & 6 \\
\hline
\end{array}
$$

Therefore one gallon contains 24 grains of Potential Ca_2OCO_2.

(28.) *Temporary and Permanent Hardness.*—The total hardness is the soap-destroying power of the water before boiling, or, indeed, before any treatment *of any kind.* The *Permanent* hardness is the hardness after an hour's boiling, there being, of course, an addition of distilled water to make up for the loss on evaporation. During boiling, bicarbonate of lime is decomposed, the carbonate being deposited, and so the water becomes softer. The

Temporary hardness is the difference between Total and Permanent hardness.

(29.)   HARDNESS OF WATER, (*examples.*)

		Ca₂OCO₂, or its equivalent in soap-destroying power.	
		Grains per Gallon.	Milligrammes per Litre.
Thames above London,	(total,)	14·0	200
New River, London,	(total,)	15·6	223
Bala Lake, . .	(total,)	0·28	4
Bala Lake, . . (permanent,)		0·21	3
Batchworth Spring Water, (Chalk,) (total,)		17·6	
Same, after being softened by Dr Clark's Process, (*vide* Clark and Smith's Paper,) . . .		2·6	

In this table of examples we have given Clark's degrees of hardness, having copied the numbers from published papers. In order to translate Clark's degrees into "*potential carbonate of lime*" we add one. Thus 14·0 is the degree given in the table for Thames water, and 15·0 is the number of grains of *potential carbonate of lime* in a gallon of Thames water. In like manner, 1·28 is the number for Bala Lake, &c. The number of grains of potential carbonate of lime is strictly proportional to the soap-destroying power of the water.

CHAPTER III.

(30.) THE determination of chlorine in water is most conveniently made by volumetric analysis. Advantage is taken of the fact that when nitrate of silver is added to a neutral solution containing chlorides and chromates the silver combines first with the chlorine, and only when the chlorine is exhausted combines with the chromic acid. Chloride of silver is white : chromate of silver is dark red.

In order to estimate the amount of chlorine in a liquid, a trace of chromate of potash is added and then standard solution of nitrate of silver dropped in until a red coloration is produced. The volume of nitrate of silver solution gives the quantity of chlorine present in the liquid.

(31.) The standard solution of nitrate of silver is made by dissolving 0·479 grms. of nitrate of silver in 1 litre of distilled water. (1 c. c. of this solution will correspond to 0·1 milligrm. of chlorine.)

(32.) The determination of chlorine in water is made as follows :

70 or 100 c. c. of water are taken, a drop of a solution containing about $\frac{1}{2}$ milligrm. of neutral chromate of potash added, and then the standard silver solution drop by drop until a permanent red colour begins to form.

If 70 c. c. of water have been taken, the number of c. c. of silver solution consumed will represent the number of *tenths of a grain* of chlorine in a gallon of water.

If 100 c. c. of water have been taken, the number of c. c. of silver solution will represent the number of parts of chlorine in one million parts of water.

The determination of chlorine made by this process is very delicate.

(33.) The importance of determining the amount of chlorine in potable water depends on the fact that the presence of an abnormal amount of chlorine points to possible sewage contamination. The quantity of chlorine present in many kinds of natural water is very small, and water which contains much of it has often derived it from sewage. It is hardly possible for water to have been contaminated with sewage without being abnormally charged with chlorides. The converse, however, does not hold, for water sometimes acquires chlorides from the geological formation through which it passes.

(34.) The following are a few examples of the amount of chlorine which is found in water. All except the first and last are taken from Watts' Dictionary :—

	Grains per gallon.		Parts in one million of water. Milligrm. per litre.
Bala Lake, (R. H. Smith,)	0·706	.	10·09
Ullswater, . .	0·693	.	9·9
Thames at Kew, . .	0·847	.	12·1
Thames at London Bridge,	4·452	.	63·6
Rhine at Basle, . .	0·105	.	1·5
Rhine at Bonn, . .	1·015	.	14·5
Elbe near Hamburg, .	2·758	.	39·4
Atlantic Ocean, . .	1330·84	.	19012·0
Well in village East of London, . .	15·61	.	223·0

CHAPTER IV.

NITROGEN EXISTING AS NITRATES AND NITRITES.

(35.) SEVERAL methods are in use for the estimation of the nitrogen existing in water in the state of nitrates and nitrites. The best method in our opinion is a modification of Schulze's, which has been recently described in the *Journal of the Chemical Society.* It consists in converting the nitrates and nitrites into ammonia by means of metallic aluminium acting upon them in the cold and in strongly alkaline solution.

(36.) The process is carried out as follows:— 100 c. c. of the water are introduced into a non-tubulated retort, and 50 to 70 c. c. of a solution of caustic soda added. The caustic soda must be free from nitrates, and the strength of the solution should be such that 1 litre contains 100 grm. of caustic soda. The contents of the retort are to be distilled until they do not exceed 100 c. c., and until no more ammonia comes over; that is, until the Nessler test is incapable of detecting ammonia in the distillate. The retort is now cooled and a piece of aluminium introduced into it, (foil will answer very well with dilute solutions, but we much prefer thin sheet aluminium in all cases.) The neck of the retort is now inclined a little upwards, and its mouth closed with a cork, through which passes the narrow end of a small tube filled with broken-up tobacco pipe, wet either with water, or better with very dilute hydrochloric acid free from ammonia. This tube need not be more than an inch and a half long, nor larger than a goose-quill. It is connected with a second tube, containing pumice stone moistened with strong sulphuric acid. This last tube serves to prevent any ammonia from the air entering the apparatus, which is allowed to stand in this way for a few hours or over night. The contents of the pipe-clay tube are now washed into the retort with a little distilled water, and the retort adapted to a condenser, the other end of which dips beneath the surface of a little distilled water free from ammonia (about 70 to 80 c. c.)* The contents of the retort are now distilled to about half their original volume; the distillate is made up to 150 c. c.; 50 c. c. of this are taken out, and the Nessler test added to them.

* Condensers are very apt to contain a trace of ammonia if they have been standing all night, and must, therefore, be washed out with the utmost care. We prefer to distil a little water through them until ammonia can be no longer detected in the distillate.

If the colour so produced is not too strong, the estimation may be made at once; if it is, the remainder of the distillate must be diluted with the requisite quantity of water.

(37.) Should it be desired to determine the ammonia by titration (alkalimetry), a much larger quantity of the water must be employed. Half a litre or a litre should be evaporated down to a small bulk and treated in exactly the same manner, except that the distillate is received in standard acid instead of water.

It will be understood that the purity of the reagents must be ascertained by making blank experiments. We have met with common caustic soda, sold in lump, which has proved to be free from nitrates. Should the caustic soda be found to contain nitrates, they may be destroyed by dissolving a small quantity of aluminium in the cold aqueous solution, and the resulting ammonia should afterwards be expelled by boiling.

The Nessler test will be described further on, under Ammonia. The process for the estimation of nitrates which we have just described is specially adapted to those cases wherein the amount of nitrates is very small, and is, we believe, the only method which gives accurate results under such circumstances.

(38.) The following examples will serve to in-

dicate the degree of trust which may be reposed
in it :—

Nitric Acid HNO₃ employed.				Ammonia NH₃ obtained.	theory.
Expt.	Milligrm.			Milligrm.	Milligram.
I.	100		(alkalimetry)	27·5	27·0
II.	100		(alkalimetry)	27·8	27·0
III.	100		(Nesslerising)	25·0	27·0
IV.	100		(Nesslerising)	28·0	27·0
V.	7		,,	1·80	1·89
VI.	1	(with salts)	,,	0·28	0·27
VII.	1	(with salts)	,,	0·26	0·27
VIII.	1	(with salts)	,,	0·25	0·27
IX.	1	(with salts)	,.	0·27	0·27
X.	1		,,	0·25	0·27
XI.	0·5		,,	0·14	0·135
XII.	·05		,.	0·12	0·135
XIII.	0·2		,,	0·05	0·054
XIV.	0·3		,,	0·078	0·081
XV.	0·1		,,	0·025	0·027
XVI.	0·1		,,	0·025	0·027

The nitric acid was taken in the form of nitrate
of potash.

In experiments VI., VII., VIII., and IX., different
inorganic salts were put into the water along with
the nitric acid in order to ascertain whether their
presence would have any effect on the reduction of
the nitric acid to ammonia. In expt. VI., 50 milligrm. of NaCl were used; in expt. VII., 50 milligrm. NaCl and 80 milligrm. K_2OSO_3; in expt.
VIII., 50 milligrm. of phosphate of soda; in expt.
IX., 50 milligrm. Fe_4Cl_6 and 50 milligrm. of

Mg_2OSO_3. It is proved that the presence of salts
has no influence on the result.

(39.) The adaptation of the method to small
quantities is very well shown by the above experiments, which exhibit it as answering just as well
with a few hundredths of a milligrm. as with a
few milligrammes of material. This extreme delicacy is due to the fact that we are able to employ
the Nessler test.

(40.) In the process, as we have just described
it, there is a *distillation*. Should there be any
reason for avoiding a distillation, the following
modification may be used :—

Prepare a soda ley by dissolving 100 grm. of
solid soda in water and diluting to a litre, and in-
sure its freedom from nitrates by taking the soda
as free from them as can be obtained in the first
instance, and subsequently dissolving a very little
aluminium in the cold solution of the alkali. Take
200 c. c. of this soda ley and mix it with 200
c. c. of the sample of water, and place in the mixture a bit of metallic aluminium. Allow the
aluminium to dissolve in the cold and the liquid to
become clear. There will thus be got a liquid containing the original ammonia and the ammonia
arising from the nitrates. Every 100 c. c. of this
liquid will contain 50 c. c. of the sample of water.

For comparison take 200 c. c. of the soda ley

and dissolve in it in the cold the same quantity of aluminium as before. When the aluminium has dissolved, add 200 c. c. of the sample of water, and allow to subside. Every 100 c. c. of the mixture will contain 50 c. c. of the water and 50 of soda ley. This *comparison-liquid* will be like the first one in every respect except one : whereas the first one has the nitrate reduced to ammonia, it will have the nitrates in the unreduced state. Decant and determine the ammonia by the Nessler test, (see § 55, using the *comparison-liquid* instead of distilled water, and using the standard ammonia of strength 0·1 milligrm. per 1 c. c.)

It will be observed that the *comparison-liquid* contains as much *original ammonia* as the liquid which is to be compared with it, and that, therefore, no correction is needed for original ammonia. Before mixing the sample of water with the soda ley—whether to make the liquid to be reduced or the *comparison-liquid*—add a very little chloride of calcium, so as to get an appreciable quantity of precipitate on mixing with the soda ley.

It will be understood that when the ammonia originally present in water is large relatively to that given by the nitrates—as occasionally happens —this plan without distillation would not be advantageous.

It is of course open to the analyst to expel the

c

free ammonia more or less completely by boiling
the water before commencing the operation of determining the nitrates.

In addition to the process we have just given,
there are Pugh's method, consisting in the reduction of nitric acid to ammonia by protochloride of
tin in sealed tubes at high temperatures, Vernon
Harcourt's reduction with iron and zinc, Pelouze's
process with proto-salt of iron, and Frankland and
Armstrong's modification of Crum's method, and a
number of other processes.

(41.) A description of Frankland and Armstrong's process is to be found in the *Journal of the
Chemical Society* for March 1868. We quote from
that journal:—

"*Estimation of Nitrogen in the form of Nitrates and
Nitrites.* . . .—The following is the mode in which this
process is applied to the estimation of nitrogen existing as
nitrates and nitrites in potable waters. The solid residue
from the half litre of water used for determination, No. 1
(estimation of total solid constituents) is treated with a
small quantity of distilled water, a very slight excess of
argentic sulphate is added to convert the chlorides present
into sulphates, and the filtered liquid is then concentrated
by evaporation in a small beaker until it is reduced in bulk
to two or three cubic centimetres. The liquid must now be
transferred to a glass tube, furnished at its upper extremity
with a cup and stopcock, previously filled with mercury at
the mercurial trough, the beaker being rinsed out once or
twice with a very small volume of recently-boiled distilled
water, and finally with pure and concentrated sulphuric
acid in somewhat greater volume than that of the concen-

trated solution and rinsings previously introduced into the tube. By a little dexterity it is easy to introduce successively the concentrated liquid, rinsings, and sulphuric acid into the tube by means of the cup and stopcock, without the admission of any trace of air. Should, however, air inadvertently gain admittance, it is easily removed by depressing the tube in the mercury trough, and then momentarily opening the stopcock. If this be done within a minute or two after the introduction of the sulphuric acid, no fear need be entertai   of the loss of nitric oxide, as the evolution of this gas does   begin until a minute or so after the violent agitati   e contents of the tube.

"Th   ixture being thus introduced, the lower extremity o e tube is to be firmly closed by the thumb, and the contents violently agitated by a simultaneous vertical and lateral movement, in such a manner that there is always an unbroken column of mercury, at least an inch long, between the acid liquid and the thumb. From the description, this manipulation may appear difficult, but in practice it is extremely simple, the acid liquid never coming in contact with the thumb. In about a minute from the commencement of the agitation a strong pressure begins to be felt against the thumb of the operator, and the mercury spirts out in minute streams, as nitric oxide gas is evolved. The escape of the metal should be gently resisted, so as to maintain a considerable excess of pressure inside the tube, and thus prevent the possibility of air gaining access to the interior during the shaking. In from three to five minutes the reaction is completed, and the nitric oxide may then be transferred to a suitable measuring apparatus, where its volume is to be determined over mercury. As half a litre of water is used for the determination, and as nitric oxide occupies exactly double the volume of the nitrogen which it contains, the volume of nitric oxide read off expresses the volume of nitrogen existing as nitrates and nitrites in one litre of water. From the number so obtained, the weight of nitrogen in these forms in 100,000 parts of water is easily calculated."

(42.) Frankland and Armstrong give two examples to show the degree of accuracy of their process. In one instance 20 milligrms. of nitre were taken, and in the other instance 10 milligrms. The results were :—

	Milligrms. of Nitrogen.	
	Calculated.	Found.
No. 1, . . .	2·772	2·897
No. 2, . . .	1·386	1·424

These determinations exhibit the me░░░ yielding results of fair per-centage accura░░d with an absolute error of $\frac{12}{100}$ milligrms. and $\frac{1}{100}$ milligrms. of nitrogen. In point of accuracy the method is probably as good as is ever actually required for water analysis. We think, however, that it is more trouble than the other method described above. Also, we do not think that it is capable of estimating quantities of nitrogen which fall below $\frac{1}{10}$ milligrm. At any rate, no proof has been offered that such small quantities are estimable by the method.

The circumstance of the binoxide of nitrogen being liberated in contact with some 8 or 10 c. c. of 50 per cent. sulphuric acid must (even though the liquid be hot) impair the accuracy of the method.

(43.) In order to give some idea of the quantities of nitrates and nitrites which occur in different kinds of water, we give a few examples taken from a table at the end of Frankland and Armstrong's paper in the *Chemical Society's Journal* (March 1868) :—

	Grains per Gallon.	Nitrogen existing as Nitrates and Nitrites. Parts per million. (Milligrms. per litre).
Thames water, as delivered in London,	0·242	3·46
River Lea water, (New River Co.,) .	0·253	3·61
Chalk water, (Kent Co.,) . .	0·286	4·08
Chalk water, (South Essex Co.,) .	0·594	8·48
Glasgow, from Loch Katrine, .	—	0·31
Manchester, from Derbyshire Hills, .	—	0·01
Lancaster, from Bleasdale Fells, .	—	0·36
Bala Lake, ...	—	0·00
Ulleswater Lake, . . .	—	0·05
Thirlmere Lake, . . .	—	0·02
Haweswater Lake, . . .	—	0·00
Well water from Leyland, near Preston, Lancashire, . .	1·726	24·66

From which it appears that whilst the London water contains comparatively much nitrates, the lake waters and the Manchester water are comparatively free from nitrates. We do not attach any value to the numbers below 0·1 in the above table.

(44.) The amount of nitrates in water has been looked on as a matter of extreme importance, and, in our opinion, unreasonably so. The nitrates and nitrites have been regarded as measuring the defilement of water. Nitrogenous organic matters decay and are oxidised, and yield more or less nitrates by the oxidation of the nitrogen which they contain.

Reflecting on this fact, some chemists have proposed to regard the amount of nitrates in a water (a small correction having been made for the nitrates existing in rain water) as the measure of the sewage which has been discharged into the water and undergone destruction. The recent reports of the Registrar-General on the state of London water refer to these data as showing "*previous sewage contamination.*"

But, if nitrates are generated by oxidation of the organic refuse discharged into water, they also find their way into it from the various geological strata traversed by the water. Chalk springs, which contain no organic matter, are often highly charged with nitrates. In this case the nitrates, if arising from sewage at all, can only be said to arise from *fossil* sewage, which was organic matter long ago in the geological ages.

Again, on the other hand, the processes of vegetation in rivers and lakes are calculated to withdraw nitrates from the water.

(45.) In fine—Presence or abundance of nitrates does not necessarily show defilement by means of sewage, and deficiency of nitrates does not show absence of defilement. And, whilst the high figure for the chalk water of the South Essex Company is no proof of previous sewage contamination of that water, the low figure of Bala Lake does not prove absence of previous sewage contamination of the water of that lake.

We do not look upon the presence of considerable quantities of nitrates in water as any bar to its employment for domestic use. We should, however, think that water of this kind would require to be exceptionally free from organic matter in order to be safe.

CHAPTER V.

AMMONIA AND ORGANIC MATTER.

(46.) WE have seen that the total solid residue left by a water, the hardness, the amount of chlorine, and the nitrates are data which in themselves are seldom of much direct value in enabling a judgment to be given whether a water is potable or not.

(47.) We now pass on to the consideration of constituents, which are of vital importance. The amount of organic matter, and closely connected with it, the amount of ammonia, is a matter of prime consequence.

(48.) The extreme minuteness of the quantity which makes the difference between a good and a bad kind of water renders this branch of the inquiry difficult, and excludes the employment of all the ordinary methods of chemical analysis. The detection and measurement of the organic matters in water belongs to the domain of micro-chemical investigation. Before describing the method which, in our opinion, really does reach the organic matters present in waters, we will discuss the other processes which have been employed for this purpose.

The method oftenest used for the estimation of organic matters in water is the determination of the loss on igniting the water-residue. The best way of getting the loss on ignition is to add a small and accurately-weighed quantity of carbonate of soda to the water before evaporating to dryness. After igniting the residue, it should be allowed to cool, and then be moistened with a solution of carbonate of ammonia, and dried first at 100° C. and then at 130° C. In this way very constant results may be got with the same water; but, unfortunately, the loss on ignition cannot be depended upon as an index to the organic matter.

(49.) Water which contains little or no organic matter sometimes contains much nitrates. Nitrates give off oxygen, and even nitrogen, on ignition, and, indeed, in the process we have given, may pass into carbonates. There may, therefore, be much loss on igniting a water-residue and still no organic matter. On the other hand, the process of evaporation to dryness is calculated to destroy organic matter, so that the loss may *under-represent* the amount of organic matter. Moreover, as we shall presently see, the ignition process is altogether of too crude an order to be suitable for dealing with the very minute quantities of organic matter which actually do occur in water.*

* Frankland and Armstrong have treated this question very

The appearance of the water-residue during ignition is probably a more valuable criterion than the
actual amount of loss. Blackening indicates organic
matter.

(50.) Instead of determining the organic matter
by the loss on ignition, it has been proposed by
Frankland and Armstrong to determine it by taking
the amount of carbonic acid and nitrogen gas which
a water-residue gives on being burnt, the nitrates
and nitrites having been destroyed beforehand by
boiling with sulphurous acid. We cannot recommend this process : it is very troublesome, and, in
our opinion, very inaccurate. A description of it is
to be found in the March number of the *Journal of
the Chemical Society*, and in the appendix to this
work will be found our remarks upon it.

(51.) Another method of measuring the organic
matter consists in ascertaining how much permanganate of potash a given volume of the water is capable
of reducing. The consumption of oxygen is taken
to be the measure of the organic matter. There are
a great many objections to this process. Nitrites
consume oxygen, and would therefore be returned
as organic matter. Different kinds of organic matter
are variously affected by the permanganate. Frank-

fully in their paper in the March number of the *Journal of the
Chemical Society*, showing that there is sometimes even *gain* of
weight on ignition of water-residue.

land and Armstrong have exhausted the objections to the use of this test in their paper (*Journal of the Chemical Society* for March 1868), and to that paper we beg to refer our readers.

(52.) The test may, however, have some little value as a rapid method of examination, where no better method is available. With very bad water, such as the Thames' water at London Bridge, permanganate is instantly decolorised, and gives very constant results when applied to the same sample of water. Dr W. Allen Miller has communicated to us a set of experiments on such water, showing the constancy of the results. With waters not so bad as that, and still not good, its indications are apt to be indecisive.

The method is practised as follows :—A solution of permanganate of potash is made by dissolving 0·320 grm. of permanganate of potash (crystals) in a litre of freshly-distilled water. The strength of the solution may be verified by noting how much oxalic acid a given volume of it will oxidize—*i.e.*, how much will be decolorised by a given quantity of the acid. The strength, as given by this prescription, is one cub. cent. = 0·08 milligrm. of active oxygen. The examinations are made by taking half a litre of the water to be examined. To this water are to be added 10 c. c. of dilute sulphuric acid, made by diluting one volume of oil of vitriol with three volumes of water. Then the solution of permanganate is added from a burette so long as it is decolorised. Different results are obtained by letting the mixture stand for different lengths of time. The temperature also affects the result. As we have said, the process is suited only for very bad specimens of water.

(53.) After having thus passed in review these methods of determining organic matter in water—methods which in our opinion are inadequate for the purpose—we pass on to a new method of determining nitrogenous organic matters. It is distinguished by its special adaptation to detect and estimate microscopic quantities, and appears to us to be especially adapted to deal with the organic impurities in water.

A very few words will suffice to enunciate the principle of this method. Most kinds of water contain ammonia, (or ammoniacal salts,) which either was recently, or will presently become, a constituent of organic matter. In addition to this, most kinds of water actually do contain more or less nitrogenous organic matter, which furnishes ammonia either on simple boiling with carbonate of soda, or else on boiling with permanganate of potash, in presence of excess of alkali.

By estimating the amount of ammonia obtainable from water, noting the circumstances under which it is obtained, we have a measure of the nitrogenous organic matter present in water.

(54.) In the whole range of chemical analysis there is no determination which surpasses that of ammonia in point of delicacy. It is questionable, indeed, whether any other approaches it. The Nessler test is capable of indicating less than $\frac{1}{200}$ milligrm. of ammonia dissolved in 100 cub. cent. of distilled water—being one part of ammonia in 20,000,000 parts of water. And ammonia admits of concentration : $\frac{1}{100}$ milligrm. of ammonia dissolved in two litres of water would, for the most part, pass into the first 100 c. c. of distillate, if the two litres of water were distilled. In this way, therefore, ammonia may be detected when the quantity is $\frac{1}{100}$ milligrm. in 2 litres of water, or 1 part of ammonia in 200,000,000 parts of water. And even this statement, surprising though it may seem, is an understatement of the delicacy of the test. Such being the character of the measurement of ammonia, the great advantage of causing determinations of organic matter to depend on measurements of ammonia will be manifest. By making these measurements of ammonia stand for measurements of organic matter in water, we apply microchemistry to water analysis.

We next treat of the measurement of ammonia by means of the Nessler test, describing the Nessler reagent :—

Measurement of Ammonia.

(55.) *Preparation of Nessler Reagent.*—This reagent is an aqueous solution of iodide of potassium, saturated with biniodide of mercury, and rendered powerfully alkaline with soda or potash.

Dissolve 50 grammes of iodide of potassium in a small quantity of hot distilled water. Place the vessel containing this solution in the water bath, and add to it a strong aqueous solution of bichloride of mercury (corrosive sublimate,) which will cause a red precipitate that disappears on shaking up the mixture. Add the solution of bichloride of mercury carefully, shaking up, as that liquid is added, so as to dissolve the precipitate as fast as it is formed. After continuing the addition of the bichloride of mercury for some time, a point will ultimately be reached at which the precipitate will cease to dissolve. When the precipitate begins to be insoluble in the liquid, stop the addition of the bichloride of mercury. Filter. Add to the filtrate 150 grammes of solid caustic soda in strong aqueous solution, (or about 200 grammes of solid potash dissolved in water.)

After having added the solution of alkali, as just

described, dilute the liquid so as to make its volume equal one litre. Add to it about 5 c. c. of a saturated aqueous solution of bichloride of mercury. Allow to subside, and decant the clear liquid, which is the Nessler reagent.

As exposure to the air is apt to render Nessler reagent somewhat turbid, it is advisable to keep the stock of the reagent in a large bottle, which should only be opened to supply a small bottle kept to hold that which is in immediate use. The addition of the 5 c. c. of solution of bichloride of mercury has two objects : It causes the Nessler reagent to clear rapidly, and imparts sensitiveness to Nessler reagent, which is deficient in sensitiveness.

(56.) *Use of the Test.*—When a small quantity of the reagent is added to a solution containing a trace of ammonia, a yellow or brown *coloration* is produced. If more ammonia is present, a *precipitate* is formed ; and if ammonia be added to the reagent, a precipitate is almost always obtained.

(57.) In order to use the test quantitatively, the following things are required :—

    (1.) Distilled water, free from ammonia.

    (2.) Standard solution of ammonia.

    (3.) A burette to measure the standard ammonia.

    (4.) A pipette for the Nessler reagent. It should deliver about $1\frac{1}{2}$ c. c.

(5.) Glass cylinders that will contain about 160 c. c.; they are graduated at 100 c. c. and at 150 c. c.

(58.) (1.) Distilled water of sufficient purity is generally to be obtained when a considerable quantity of water is distilled. The first portions of distilled water usually contain ammonia. After a while, on continuing the distillation, the water usually distils over in a state of tolerable purity, but towards the end ammonia will again appear in the distillate. By collecting the middle portion of the distillate apart from the rest, it will usually be easy to obtain distilled water of sufficient purity. In order to be available, the distilled water should not contain so much as $\frac{1}{100}$ of a milligrm. of ammonia in 100 c. c. of water. If there is no opportunity of distilling a large quantity of water, and taking the middle fraction of the distillate, it may be necessary to re-distil distilled water of ordinary quality : the first part of the distillate will be ammoniacal, after that there will be water free from ammonia.

(59.) (2.) The standard ammonia should contain $\frac{1}{100}$ of a milligrm. of ammonia in one cubic centimetre of water. It is made by dissolving 0·03882 grm. of sulphate of ammonia in a litre of water. If chloride of ammonium be taken, the quantity of the chloride to be dissolved in a litre of water is 0·0315 grm.　It will be found most convenient in practice to keep a solution of ten times this strength, (0·3882 grm. sulphate of ammonia in a litre of water,) and to dilute it when required for use.

(60.) In order to estimate ammonia, fill one of the cylinders up to 100 c. c. with the solution to be examined, and add $1\frac{1}{2}$ c. c. of Nessler reagent by means of the pipette.　Observe the colour, and then run as much of the standard solution of ammonia as may be judged to correspond to it into another cylinder containing distilled water, and fill up with water to 100 c. c.　Allow the liquids to stand for ten minutes.　If the coloration is equal, the amount of standard ammonia used will represent the ammonia in the fluid under examination.　If not, another cylinder must be filled, employing a different amount of the standard ammonia, and this must be repeated until the colours correspond.　It is very seldom necessary to make more than two such comparative experiments ; and with a little practice, the operation of Nesslerizing will become very easy and rapid.　With regard to the limits of the readings, it is not difficult to recognise $\frac{1}{100}$ of a milligrm. of ammonia in 100 c. c. of water, and the difference between $\frac{19}{100}$ and $\frac{20}{100}$ of a milligrm. should be visible. It will be observed that $\frac{1}{200}$ milligrm. of ammonia will be more visible in 50 c. c. of water than in 100

D

c. c. of water, so that when it is desirable to detect
the very minutest quantities, concentration of NH_3
in a small bulk of water is to be recommended.

With regard to the superior limit. When the
ammonia becomes too concentrated, precipitation
occurs. Different samples of Nessler test will sus-
tain different quantities of NH_3 without precipitation.

The presence of a great number of substances in
aqueous solution containing ammonia will interfere
with the indication of the Nessler reagent, and it is
always desirable to have the ammonia in pure distilled water, if that be possible. In order to do so,
the solution containing the ammonia should be distilled with a little alkali, and the Nessler reagent
applied to the distillate.

(61.) When there is a necessity for the use of
the Nessler test without previous distillation, a
special device has to be resorted to in order to get
rid of the disturbing influence on the Nessler test
of the substances dissolved. This device will be
explained further on.

Ammonia-method of Water Analysis.

(62.) The examination of water by the new
method, which we may characterise as the ammonia-
method, is managed as follows:

Half a litre of water is taken and placed in a

tubulated retort, and 15 c. c. of a saturated solution of carbonate of soda added. The water is then distilled until the distillate begins to come over free from ammonia, (*i.e.*, until 50 c. c. of distillate contain less than $\frac{1}{100}$ of a milligrm. of NH_3. A solution of potash and permanganate of potash is next added. This solution is made by dissolving 200 grms. of solid caustic potash and 8 grms. of crystallised permanganate of potash in a litre of water. The solution is boiled to expel any ammonia, and both it and the solution of carbonate of soda ought to be tested on a sample of pure water before being used in the examination of water.

50 c. c. of this solution of potash and permanganate should be used with half a litre of the water to be tested.

The distillation is continued until 50 c. c. of distillate contain less than $\frac{1}{100}$ milligrm. of ammonia.

Both sets of distillate have the ammonia in them determined by means of the Nessler test as described above.

No matter how good the water may be, it is desirable never to distil over less than 100 c. c. with carbonate of soda, and not less than 200 c. c. after the addition of the potash and permanganate of potash.

(63.) It will easily be understood that the greatest

cleanliness is requisite in carrying out this process.
The Liebig's condenser is especially liable to contain
traces of ammonia, and should be cleaned out im-
mediately before being used. The best way of effecting this is by distilling a little water through it.
In boiling the contents of the retort it is well to
use the naked flame placed *quite close* to the retort,
so as not to heat one spot only. We are in the
habit of using a large Bunsen-burner placed close to
the bottom of the retort. Persons who are not in
the habit of distilling with the naked flame will probably find an argand gas-lamp with a metallic chim-
ney to be the more convenient source of heat.

With regard to the retort itself, it should be
capable of holding about 1500 c. c. when in position for distillation and filled up, so as to run over.
The tubulure should be so situated as to admit of
the charge being poured in when the retort is *in
situ* for distillation. The charge is to be introduced
by means of a funnel so as to avoid dirtying or
cracking the retort.

Longevity of a good retort.—Our experience is
that a retort either breaks the first time or two of
using, or else lasts 50 or 100 times. One retort
performed 217 operations.

(64.) The following examples will serve to illustrate the process :—

Edinburgh water (from Swanston), 25th September 1867.

ONE-HALF LITRE TAKEN.

	c. c.		Milligrm. NH₃
Distillate I. (carbonate of soda),	100	=	·015
Distillate II. (potash and permanganate of potash), .	100	=	·035
	100	=	·015
			·050

Therefore, 1 litre of Edinburgh water (from Swanston) contains :—

			Milligrm.
Free ammonia,	.	.	0·030
Albuminoid ammonia,	.		0·10

or 1,000,000 parts contain :—

Free NH₃,	.	.	.	0·03 parts.
Albuminoid NH₃,	.	.		0·10 parts.

Glasgow water (Loch Katrine) October 3rd, 1867.

ONE-HALF LITRE TAKEN.

	c. c.		Milligrm. NH₃ ?
I. Distillate (carbonate of soda),	100	=	·002
II. Distillate (potash and permanganate of potash), .	100	=	·04
	100	=	·00
			·04

Therefore one litre contains :—

			Milligrm.
Free NH₃	.	.	.	0·004
Albuminoid NH₃ ; .	.		0·08

London water (Thames) West Middlesex water from the main, 25th January 1868 :—

ONE-HALF LITRE TAKEN.

	c. c.		Milligrm. of NH_3
I. Distillate (carbonate of soda),	100	=	·02
II. Distillate (potash and permanganate of potash), .	100	=	·07
	100	=	·01
			——
			·08

Therefore 1 litre contains :—

			Milligrm.
Free NH_3,	.	.	0·04
Albuminoid NH_3,	.	.	0·16

(65.) Sometimes it is necessary to add distilled water to the contents of the retort. This happens in the case of very bad waters which require prolonged boiling in order to effect complete decomposition of the organic matters. Whenever this has to be done the greatest care must be taken to add only pure water.*

(66.) As has been already mentioned, the first portion of ammoniacal distillate contains both the free ammonia of the water and that obtained from the decomposition of any urea that may exist in the

* The distilled water to be used for this purpose should not only refuse to react on Nessler reagent, but must contain no organic nitrogenous matter. It must give no ammonia on distillation with alkaline permanganate.

Bad water often bumps much during the boiling with the alkaline permanganate. When this happens it is well to use fragments of broken-up tobacco pipe previously ignited, which will stop bumping.

water. Usually it is quite unnecessary to make any separation of the free ammonia from the ammonia present as urea. In the case, however, of very foul water, as for instance in the Thames water taken at London Bridge, it is sometimes worth while to make this distinction. When this is desired, a determination of the free ammonia actually present in the water must be made. The difference between the amount of ammonia evolved by carbonate of soda and the ammonia present as such is equal to the ammonia obtained from the urea.

(67.) There are some difficulties in estimating the ammonia, present as such, in water. As we have already remarked the presence of many substances interferes with the Nessler test, and we have to elude this difficulty. This may be done by operating as follows :—

Take 500 c. c. of water, add a few drops of solution of chloride of calcium, then a slight excess of potash. Filter. Put it into a retort and distil until the distillate comes over, free from ammonia, then make up the contents of the retort with distilled water to their original volume, viz., 500 c. c. Now take 200 c. c. of the original water, treat it with chloride of calcium and potash as before. Then filter, care being taken to have the filter paper well washed before commencing the filtration. In this way two samples of water are obtained, the one, with the am-

monia as in the original water, and the other without the ammonia, but in every other respect the
same as the former. This second portion of water
is to be used in the place of distilled water to make
the Nessler comparisons; as both samples contain the
same impurities, they will affect the tint of the
Nessler test in the same manner. Thus the error
arising from the presence of salts, &c., is avoided.

(68.) *Example.*—A London well water which
contained 192 grains of solid matter per gallon, (or
2743 in the million) was taken. It was boiled with
a little potash, (the addition of chloride of calcium
being unnecessary) until it was free from ammonia.
It was then diluted to its original volume with pure
distilled water, 100 c. c. of this diluted water were
taken, and $\frac{8}{100}$ of a milligrm. NH^3 added to it, and
then the Nessler test. There was no colour at first,
but after a little while it became coloured, but the
colour never equalled that given by $\frac{8}{100}$ of ammonia
in the same volume of distilled water.

In another case the ammonia was estimated in
a well water, the comparisons being made, both
with distilled water and with some of the water
prepared as above. The results were as follow:—

	Milligrm. NH^3 per litre.
Comparison with well water,	·75
Distilled water,	·60

In this case also the total solid residue was large,

·208 grains per gallon, or 2971 parts in the million. Other examples might be given, but all that it is necessary to say is, that the error appears to be sometimes on the other side.

(69.) We now proceed to give the results furnished by different waters which we have examined. Under "*free NH₃,*" should be understood NH_3 driven out by carbonate of soda. Under "*Albuminoid NH₃*" is given the NH_3 evolved by boiling with alkaline-permanganate, as was described.

(69-a.) London water from the *Thames*, as supplied in London by the *different companies*.

Parts in 1,000,000 (*milligrm. in a litre.*)

Date.	Name of Company.	Free NH₃.	Albuminoid NH₃.
1867.			
July	West Middlesex Co.	·01	·066
,,	,, ,,	·01	·064
,,	,, ,,	·01	·064
,,	,, ,,	·01	·06
,,	Grand Junction Co.	·01	·08
,,	,, ,,	·01	·07
,,	,, ,,	·01	·07
,,	,, ,,	·01	·07
,,	Chelsea Co.	·01	·07
,,	,,	·01	·10
,, 6.	Southwark and Vauxhall Co.	·03	·16
,, 18.	,, ,,	·015	·12
,, 18.	,, ,,	·01	·15
,, 19.	Lambeth Co.	·015	·14
,, 20.	,,	·015	·15

The different determinations of the same company's water were made on samples taken from different parts of the district.

(69-b.) London Water Supply.—*The New River Company.*

Parts in 1,000,000 (*milligrm. in a litre.*)

Date.		NH₃ Free.	NH₃ Albuminoid.
1867.			
June 21.	Cab-rank, Moorgate Street,	·015	·084
July 12.	London Institution,	·010	·05
,, 20.	Supply-pipe in house, Kingsland Road,	·015	·05
Aug. 21.	Cab-rank, Moorgate Street,	·002	·064
1868.			
Jan. 25.	London Institution,	·02	·09

(69-c.) London Water Supply.—*River Lea.*

1867.			
June 18.	East London Water Co.	·03	·089

(70.) Manchester Water Supply.

1867.			
Aug. 19.	Brazennose Street,	·006	·06
,,	,, ,,	·01	·07
,,	Cateaton Street,	·014	·07
,,	Palatine Hotel,*	·014	·10

(71.) Edinburgh Water Supply.

1867.			
Sept. 25.	Swanston,	·03	·10
,,	Crawley,	·006	·08
,,	Colinton,	·14	·08
,,	Comiston,	·004	·034
Sept. 18 and 19.	⌈ Edinburgh supply ⌉	·00	·075
,, ,,	⎱ from the Labo- ⎰	·00	·070
,, ,,	⎰ ratory of the ⎱	·00	·063
Oct. 4.	⌊ University, ⌋	·004	·100

* Probably slightly contaminated.

(72.) Glasgow Water Supply (*Loch Katrine.*)

Date. 1867.	NH₃ Free.	NH₃ Albuminoid.
Oct. 3.	·004	·08

(73.) The Edinburgh water supply consists of four kinds of water—Swanston, Crawley, Colinton, and Comiston. They are mixed in a reservoir, and the water sent to the city of Edinburgh is the mixture. As will be observed, the Comiston water was excellent at the time the examination was made.

On examining the above table, it will at once be apparent that there was a markèd difference between the water as supplied by the different Thames' companies last summer. Thus, whilst in July last the West Middlesex and Grand Junction companies were supplying water containing ·06 milligrm. to ·08 milligrm. of albuminoid ammonia per litre, the Southwark and Vauxhall Co. supplied ·12, ·15, and ·16 milligrm. per litre. Frequent examinations of the Southwark and Vauxhall Co.'s water, conducted at different periods of the year, has confirmed this pre-eminence in badness for that company. In short, the water supplied by this company has at times been almost on a par with unfiltered Thames' water as it exists in the river at Hampton Court.

(74.) Taking, however, the water of the West Middlesex Co. and the Grand Junction Co. as properly filtered samples of Thames water, we shall, on looking through the table, see an extraordinary regularity in the albuminoid character of good town waters during the summer months. Thus :—

During Summer Months.

	Albuminoid NH$_3$ per litre milligrm.
Thames (well filtered by Co.'s,) .	·06 to ·08
New River (London,) . .	·05 to ·064
Manchester, ...	·06 to ·07
Edinburgh (town supply,) . .	·063 to ·075
Glasgow (Loch Katrine,) . .	·08

The great constancy of these numbers is a most remarkable thing, especially when it is borne in mind how very different are the sources of the waters which yield them. The Thames water, as it exists in the river near Hampton Court, has received much refuse of various kinds, even sewage. The Manchester water comes off the moorlands of Derbyshire, being collected in reservoirs at Woodhead, and conveyed thence in pipes to Manchester. The Edinburgh supply comes partly from springs, which are carefully tended, and which arise some miles from that city. The Glasgow water comes from Loch Katrine. And yet, notwithstanding this diversity of origin and of history, we find the character of all of them so much alike that the albuminoid ammonia is comprised between the narrow limits ·05 to ·08 parts. It would seem as if there were natural processes at work, tending to equalise the quality of natural water that is freely exposed to air and light.

(75.) A circumstance of some importance, to which we shall now direct attention, is the great improvement effected in water by filtration, as illustrated by the difference in character between Thames water near Hampton Court and Thames water as supplied by some of the water companies.

(76.) The following results were given by Thames water, taken from the river, near Hampton Court, on 9th July 1867 :—

Parts in the 1,000,000.

	Free NH3.	Albuminoid NH3.
I. Water taken above the Weir, and some distance above Hampton Court, ...	0·045	0·28
II. Another sample taken below the Weir, ...	0·015	0·23
Sample I., filtered through filter paper,	0·045	0-21
Sample II., „ „	0·015	0·185

In July 1867 this same Thames water, after the very excellent filtration effected by certain water companies, was supplied in London, having the character indicated by the numbers,—Free NH3, 0·01 ; Albuminoid NH3, 0·06.

(77.) Experiments have been made to ascertain the degree of organic purification effected by Clark's admirable softening process, viz., by the addition of lime-water to a hard water, so as to throw down carbonate of lime. The results, as will be seen, are very striking, and tend in the

same direction as the observations on the state of
the Thames before and after the efficient filtration
effected by the water companies. This table will,
we think, speak for itself :—

Parts per 1,000,000.

		Free NH$_3$.	Albuminoid NH$_3$.
I.	Before Clark's process,	0·01	0·05
	After „	0·01	0·02
II.	Before Clark's process,	0·025	0·22
	After „	0·030	0·08
III.	Before Clark's process,	0·015	0·22
	After „	0·020	0·07
IV.	Before Clark's process,	0·195	0·12
	After „	0·15	0·06

It appears, moreover, that the organic matter
carried down by the carbonate of lime, and so
removed from the water, is more highly organised
than that which remains behind in solution. (Un-
published experiments by Mr Chapman.)

All drinking waters may therefore reasonably be
required to be of such a degree of purity as not to
yield more than 0·08 milligrm. of albuminoid
ammonia per litre of water.* If not in this state,
naturally, all water that is to be used for drinking
ought to be filtered till it becomes so.

(78.) We will now present some specimens of the

---

* We have, we think, insisted upon less than might be required. Possibly a very high degree of purity, say ·03 milligrm.
albuminoid ammonia per litre will be generally attainable
without unreasonable trouble.

results given by waters of unquestionably bad character, so as to provide the water-analyst with the typical characteristics of foulness :—

Parts in 1,000,000 (milligrm. in a litre.)

Date. 1867.		NH₃ Free.	NH₃ Albuminoid.
June 21.	Great St Helen's Pump, London, . . .	3·75	0·18
„ 28.	Pump in Bishopsgate Street, London, . . ·	7·50	0·255
„ 29.	Pump in Draper's Hall, London, . . .	6·00	0·315
„ 18.	River Thames. Midstream at London Bridge tide, two hours' flood. After filtration through filter paper. . . .	1·76	0·35
„ 18.	Thames. London Bridge. High Tide, . .	1·02	0·59
	Do., do., .		0·56
	Do., do., .		0·50
Water from a pump in Edinburgh, .		0·21	0·29
Water from a well in a village east of London, after heavy rain, .		40·00	3·00

The water from this last well contained 2240 parts of solid residue per million of water, and 223 parts of chlorine per million.

(79.) Now, the quantities per litre indicated by the above figures, although most striking, and quite unmistakable when displayed in coloration of Nessler test, are nevertheless small quantities when considered in themselves.

Between $\frac{59}{100}$ of a milligrm. of albuminoid ammonia, given by a litre of dirty water from the Thames at London Bridge; and $\frac{8}{100}$ of a milligrm. of NH, per litre of New River water. This is an example of the difference to be met with as distinguishing between water that is vile and stinking, and water that is wholesome.

(80.) Miscellaneous Examples :—

1867.		Free NH³	Albuminoid NH₃
July ...	Bala Lake water . .	·01	·25
June...	Wolsey's well, Wimbledon .	·03	·159
„	Water from Cold Harbour, Dorking . . .	·015	·00
„	Caterham water, after Clark's process . . .	·04	·00

DETERMINATION OF UREA.

(81.) In the following instances the total ammonia was determined by distillation. For example :— 1 litre of the water was put into a retort, and about 30 grammes of dry carbonate of soda, dissolved in a little water, added; the retort was adapted airtight to a condenser, the other end of which dipped beneath a little distilled water contained in a vessel large enough to hold about 600 c. c. The retort was placed on a sand bath, and gently heated until the water just commenced to simmer. The distillation was carried on very slowly until about half the contents of the retort had distilled over. The time taken by this distillation was about five hours.

The receiver was now removed, and another put in its place, the distillation being quickened. After about 100 c. c. had come over, they were tested for ammonia; and if found free, which they for the most part were, the process was at an end. If not, then the ammonia was estimated in the 100 c. c., and a note taken of the quantity. The distillation was now continued as before, until the distillate was practically free from ammonia. Great care must be taken in these distillations not to let the heat play on the incrustation which forms above the liquid, as almost all forms of nitrogenous organic matter will, when heated in this way, give off ammonia. In all cases, the first 500 c. c. contained the great bulk of the ammonia, about 90 to 95 per cent. at least, and in three out of the four examples it contained almost all.

(82.) The estimations of ammonia without distillation were made as described elsewhere in this book.

Well-water known to be very much contaminated with fresh sewage :—

	Milligrammes per litre.
Ammonia by distillation . . .	20·5
Ammonia without distillation . . .	18·0
Difference due to urea or kindred bodies	2·5

which would indicate 4·4 parts of urea per 1,000,000.

Ammonia by distillation . . . 11
Without 10
—
Therefore from urea . . . 1

Example on known quantities :—

Hard water, almost free from organic matter and
ammonia; added 3 milligrammes per litre urea,
10 milligrammes ammonia, also about 1·6 milligrammes of gelatine. The gelatine had been made
up into solution about three weeks before the experiment, but it had very little smell.

Ammonia by distillation . . . . 12·5
Without distillation . . . . . 10·5
—
2·

Difference due to urea, and therefore—
Urea found . . 3·5
Theory . . . 3·0

Same experiment, substituting albumen for gelatine. The exact quantity of albumen was not
known; it could not have been far off the quantity
of gelatine; the albumen (white of egg) had been
thirteen days in concentrated aqueous solution.

Ammonia by distillation, over 12 and
under 12·5 ; say 12.25
Without distillation . . . . 10·5
—
1·75

Therefore urea, 3·1
Theory,   .   3·

This last example must be regarded as much within the ordinary limits of error attendant on such determinations, for the per-centage of error on both the large quantities of ammonia must fall on the ammonia from urea. Still, by taking care always to keep the estimations slightly in excess, or always slightly in deficit, the errors may be made to some extent to neutralise each other.

(83.) *Interpretation of the "Albuminoid Ammonia :"—*

Having shown the amounts of "albuminoid ammonia" obtainable from natural water of different kinds, we have next to consider the meaning of this datum.

An extended series of experiments made upon the action of alkaline permanganate on nitrogenous organic substances of different kinds (care being taken to include the utmost variety of chemical character in the examples taken for experiment) has brought to light the fact that nitrogenous organic substances in general, when they do not contain the nitrogen in the nitro-state or as urea, evolve it as ammonia. A large class of substances evolve all the nitrogen as ammonia; other substances evolve only a part of it as ammonia.

The following are examples of substances which give up all their nitrogen as ammonia :—

Asparagine, ...	22·66
Piperine, ...	5·96
Diamylamine Chloride, . . .	8·79
Amylamine, ...	19·54
Diphenyl-tartramide, . . .	11·33
Piperidine, Sulphate, . . .	12·69
Hippuric acid, ...	9·50
Narcotine, ...	4·11

The following substances have yielded the half of their nitrogen as ammonia :—

	Theory NH$_3$.	Found NH$_3$.
Morphine, ...	2·98	2·80
Codeine, ...	2·67	3·00
Papaverine, . . .	2·50	2·20
Strychnine, . . .	5·09	5·45
Iodide of methyl-strychnine, .	3·57	3·33
Brucine, ...	4·32	4·60
Sulphate of quinine, . .	4·56	4·50
Sulphate of cinchonine, . .	4·76	{ 5·7 { 5·4
Nicotine, ...	10·49	10·80
Naphthylamine, . . .	5·95	{ 6·65 { 6·81
Toluidine, . . .	7·95	{ 8·83 { 8·30
Acetate of Rosaniline, . .	7·06	{ 6·37 { 6·49

The numbers express the amount of ammonia given by 100 parts of the substance.

100 parts of gelatine have given 12·7 parts of NH$_3$.
100 parts of caseine* gave 7·6 parts of NH$_3$.

* Of doubtful purity.

100 parts of dry albumen give about 10 parts of NH_3.

100 parts of uric acid have given about 7 parts of ammonia.

100 parts of creatine have given 12·6 parts of ammonia.

100 parts of theine have given 8·54 parts of ammonia.

The experiments were made on microscopic quantities of substance.*

It has also been shown, that very dilute solutions of albumen having been made, and operated upon, the ammonia obtained is strictly proportional to the amount of albumen employed. Lastly, experiments have been made on gluten, horn, hairs, fur, spiders' eggs, and a variety of animal substances, and all of them give off ammonia on being heated with alkaline permanganate. With the aid of these experimental materials we are prepared to deal with the datum, the "albuminoid ammonia" of waters. Let us, first, treat the question purely in the abstract, and, disregarding the history of the water in the Thames at London Bridge, inquire what amount of nitrogenous substance may this Thames water possibly contain. A litre of an aqueous solution of organic matter yields $\frac{52}{100}$ milligrm. of ammonia on distillation with alkaline permanganate, how much nitrogenous organic substance (not nitro-compounds) can it possibly contain?

* See *Journal of the Chemical Society*, May 1868.

(84.) Of the great number and variety of nitrogenous substances tried by us the one which gave the lowest per-centage of albuminoid ammonia was papaverine—*i.e.*, 2·2 per cent.

The highest quantity of nitrogenous organic substance possible in a litre of this Thames water would be accordingly 27 milligrms.

Suppose the albuminoid ammonia came from strychnine, the quantity of strychnine would be 11 milligrammes in the litre of water, which would thus contain about a medicinal dose of that alkaloid.

Suppose the albuminoid ammonia to proceed from a substance rich in nitrogen, and capable of evolving all the nitrogen as ammonia, from asparagine, the most extreme case on our list, from which 22·7 per cent. of albuminoid ammonia are to be obtained. In this case, the litre of water would contain about 2·5 milligrms. of nitrogenous organic substance.

Thus, regarded purely in the abstract, this litre of water, having been deprived of ammonia and any ureas possibly present, furnished 0·59 milligramme of ammonia on distillation with alkaline permanganate. It contained, therefore, at the maximum, 27 milligrammes of organic nitrogenous matter, and at the minimum, 2·5 milligrammes.

In like manner, the New River water which has given 0·05 milligrm. of NH, per litre of water, if we knew absolutely nothing about its possible constituents, could be set down as containing in maximum about 2·4 milligrm., and in minimum 0·2 milligrm. of nitrogenous organic matter. If the substance in New River water were strychnine, one litre of the water would contain rather less than 1 milligrm. of that alkaloid.

(85.) But, when we make use of the facts known concerning the history of natural waters, then the information afforded by a determination of the amount of albuminoid ammonia which they furnish becomes much more explicit.

The Thames water at London Bridge derives the main part of its organic matter from sewage—from animal and vegetable refuse. Most of the nitrogenous organic substance which has escaped decomposition, and which furnishes the "albuminoid ammonia" is doubtless either albumen, or closely allied to that substance. Ten parts of "albuminoid ammonia" are given by 100 parts of albumen. The nitrogenous organic matter in this water is, therefore, about ten times the weight of the albuminoid ammonia, and at this rate a litre of this Thames water will contain about 6 milligrm. of nitrogenous organic matter.

The products of the metamorphosis of albumen
and animal tissues, generally such as :—

Uric acid, .	.	7·	parts NH_3 per 100	
Creatine, .	.	12·6	,, ,,	,,
Hippuric acid,	.	9·5	,, ,,	,,
Gelatine, .	.	12·7	,, ,,	,,

may be expected to form a mixture which would
evolve about $\frac{1}{10}$ of its weight of "albuminoid ammonia;" so that even the disintegrating animal
refuse in the river would be pretty fairly measured
by ten times the albuminoid ammonia which it
yields.

CHAPTER VI.

METALS.

(86.) UNDER this head we shall only consider in relation to the subject of water analysis the heavy metals, arsenic, lead, copper, zinc, iron, and manganese. Chemists are in the habit of neglecting to examine water for the heavy metals, and yet the presence of these metals may reasonably be suspected in many of the sources which have been proposed for the supply of towns, and metallic compounds are very active physiologically. The Cumberland water, for instance, is all of it to be suspected of containing these metals, and examination would doubtless disclose the presence of traces of them.

(87.) Owing to the absence of investigation we are quite in the dark as to the effect of very minute traces of these metals on the health of people who take them habitually in drinking water; and in writing this chapter we are reduced to the necessity of pleading for the collection of data rather than of laying down what is to be concluded from the presence of a certain minute quantity of such and such a metal.

It will be admitted on all hands that the presence of any considerable quantity of any one of the first four—say a grain of metal in a gallon of water would constitute a bar to the employment of that water for domestic use.

(88.) There is, however, a whole field of investigation all untouched in the physiological action of very minute traces of different metals existing in water. Possibly a part of the sanitary effect of what is called "change of air" is due to change in the minute metallic impurity in the drinking water of the parts of the country which are visited. The effect of metallic compounds taken internally with the food or drink is cumulative. The quantity of water consumed by one individual in a week is very large. Very minute quantities of a metal in drinking water may, for aught we know to the contrary, determine the sanitary character of a district. It is notorious that mountainous parts of a country are not healthy as places of constant, abode, and the water in mountainous districts is specially liable to contain traces of the poisonous metals.

The quantity of water drunk in various shapes during the space of twenty-four hours by an adult man may be set down on the average as $1\frac{1}{2}$ litres.

The quantity is of course very variable, but this figure, which is rather below the mark, will be a sufficiently good approximation for our present purpose. This would be at the rate of ten litres a week. If, then, there were only one milligramme of arsenic in a litre of the drinking water, there would be twenty milligrammes of that metal taken in the space of a fortnight. Twenty milligrammes of arsenious acid taken at once has been known to be fatal. Extending over longer periods of time —in illustration of the effect on the natives and on the inhabitants of a metallic district— $\frac{1}{100}$ milligrammes of lead in each litre of the drinking water would mount up to about $5\frac{1}{2}$ milligrammes of lead as the annual consumption.

We are all in the dark as to the length of time during which the animal economy would absorb these traces of metallic compounds from drinking water. Granted that a man took $\frac{1}{100}$ milligramme of lead daily in his drink, and continued to do so for years, how long would his system go on absorbing and accumulating the poison ? How much would be absorbed in a given time ? Is there a process of excretion of such poisons parallel or partly parallel with their absorption ? Is dilution beyond a certain limit a bar to absorption by the animal economy ?

On most of these points we are very much in the

dark. But, if analogy be allowed to guide us, and if we may trust general laws, we may return an answer to the last query. The limit to the dilution in which a metallic compound dissolved in drinking water is capable of being absorbed, is very far off.

(89.) From the beginning to the end of life we are, along with our other avocations, engaged in performing a most careful evaporation of part of the water that we drink. The liquid goes in at the mouth, and having passed into the stomach and intestines, undergoes a miraculous filtration in passing through the walls of the capillary vessels into the blood. Having thus been filtered and had extraction performed upon it once by the walls of the capillaries, on its entrance it again undergoes extraction on its way out of the system. This extraction occurs either as the liquid passes through the structure of the kidneys into the ureters, whence it proceeds into the bladder and out of the body in the shape of urine; or in the lungs, and in the sweat glands as the liquid again passes through walls of tissue. In this last case there is, in addition to extraction, much actual evaporation of the most careful kind. Water escapes from the lungs and the skin in the form of vapour.

(90.) Conceive an evaporating basin covered in with many folds of bibulous paper and kept in a draft of air of about 30° C. and kept constantly fed with liquid and so suffering continuous evaporation of the most careful kind. Conceive this to go on for seventy years, and you have a picture of the kind of process that we are carrying on in our bodies. We evaporate daily through the lungs and skin, say 500 grammes of water. In twenty years we evaporate therefore 3650 litres. And this is evaporation at 30° C. and through tissues. No matter how minute the metallic impurity, this evaporation must cause it to accumulate in the animal tissues.

(91.) These considerations enable us to assign a limit to the dilution of a metallic poison and to specify a point of dilution beyond which a metallic poison must become innocuous.

Taking the evaporation as the basis of the calculation, we have seen that in twenty years there is an evaporation of about 3650 litres. In 60 years there would therefore be 10,000 litres evaporated.

A water, which does not contain a poisonous dose of metallic poison in 10,000 litres, is certainly, metallically considered, a good water.

Or, if we add the accumulating power through extraction of metallic compounds, performed by the animal tissues, to the concentration effected by evaporation ; in short, if we suppose all the water drunk during a lifetime to have all its metallic salts extracted from it by the body of the animal which

drinks it, we are certainly beyond the limit. At
this rate, when 30,000 litres of the water contain
less than a poisonous dose, the water cannot possibly be bad, metallically considered.

We are thus enabled to fix a limit to the poisonous
quality of metallic impurity. *Water, which does not
contain so much as $\frac{1}{2000}$ milligramme of metallic
poison per litre*, is, undoubtedly, inoperative, metallically considered.

(92.) How far, within these limits, a water is
metallically safe, has yet to be ascertained. We
should be inclined to suspect that $\frac{1}{100}$ milligramme
of certain metals, per litre of water, may produce
effects on the general health of the community which
drink the water.

In treating of ammonia, we explained that less
than the $\frac{1}{100}$ of a milligramme in a litre of water
was capable of being detected by the Nessler test, a
preliminary concentration having been made, so as
to obtain the ammonia in the distillate. If the task
were the detection of metals present in minute
quantities of this order of minuteness, we should
have to concentrate, and should usually concentrate,
so as to get the metal to accumulate in the residue,
not in the distillate, as in the case of ammonia.

(93.) The general method of accumulating metals
is to take a large volume of water—say 5 or 10
litres—render it alkaline with a little potash or soda,

and evaporate down to a very small bulk, which is to be variously treated, according to circumstances.

We hope, at some future date, to be in possession of special methods for the estimation of minute traces of metals in water. On the present occasion, we can do little more than just indicate the methods of detection which promise well.

(94.) *Arsenic.*—Having obtained the small bulk of concentrated water, render it acid with pure hydrochloric acid, and introduce it into a Marsh's apparatus, which is evolving pure hydrogen, and look out for the arsenical spot on porcelain, and the deposit of metallic arsenic in the interior of the heated tube. From the extent of the metallic deposit, some idea of the quantity of arsenic may be formed.

(95.) *Lead* and *Copper.*—Having lightly acidulated the concentrated water, add a little sulphuretted water, and look for blackening of the liquid. Blackening or darkening of the liquid points to the presence of these and certain other metals, and is very easily seen against a white ground.

(96.) *Zinc.*—We cannot undertake to say which of the various possible methods will be suitable for the extremely minute quantities to be looked for. There is, however, hardly any metal which deserves to be more carefully sought in water than does zinc.

Manganese.—This metal sometimes occurs in water. We have excellent minute tests for it in the formation of manganate of soda before the blow-pipe, and also in the action of solution of peroxide of hydrogen on the solution of it which has been rendered alkaline. · A very bulky brown precipitate is given by peroxide of hydrogen in extraordinarily dilute solutions of manganese salts. When the solution is very dilute, the reagent gives a brown coloration. We believe Dr Angus Smith first pointed out the great advantages of peroxide of hydrogen as a test for this metal.

(97.) *Iron.*—The presence of a trace of iron in water cannot be pronounced to be an objection to its domestic use. But a water ought not to contain much of this metal. The best way of estimating it is volumetrically, by reducing its solution to the state of solution of proto-salt of iron. This may be done with sulphurous acid, the excess of which is afterwards driven off by boiling. The amount of iron is then ascertained by noting how much per-manganate of potash is decolorised by the solution.

(98.) It is well to test water for baryta-salts, for these sometimes occur in very unexpected circum-stances.

(99.) In fine, we would point out to the water analyst the desirableness of paying attention to the minute traces of heavy metals present in so many kinds of natural water. From a sanitary point of view, these traces are much more important than the exact determination of the relative amounts of lime, magnesia, silica, alumina, and alkalis in a water-residue.

(100.) As an example of the occurrence of minute quantities of metals in natural water, we may refer to the instance of the town of Cheddar, in the west of England, where a portion of the ordinary water supply is distinctly poisonous, and was found to contain lead. This part of the water supply has ceased to be used for domestic purposes.

Copper has also been recognised as occurring in natural waters. Sir Samuel Baker mentions the existence of a stream near Abyssinia, which, during the months of drought, became very decidedly impregnated with copper. In the seasons of flood, the water was considered by the natives to be good. During the dry season, the natives suffer much from the effects of the copper. Instances of this kind are common in mining districts.

APPENDIX.

As we have omitted to give Frankland and Armstrong's method of determining organic matter in water, we think it necessary to give our reasons for so doing. They may be gathered from the objections raised to it in the following note, reprinted from the *Journal of the Chemical Society*, April 1868 :—

In the memoir lately published[*] by Frankland and Armstrong, describing the method by which they propose to conduct the examination of waters as regards organic substance, objections are raised to the method we have recommended for determining the relative quality of water by means of the albuminoid ammonia it yields by treatment with alkaline solution of permanganate of potash. Those objections are, to a great extent, based upon a comparison of the results obtained in this way with the results obtained by Frankland and Armstrong's method. We have, therefore, made a careful inquiry

[*] *Chem. Soc. Journal*, March 1868.

into the conditions to which the difference between the results furnished by the two methods may be referred, and into the adequacy of Frankland and Armstrong's method to be applied as a test of the results furnished by our method.

In placing before the Society the conclusions at which we have arrived, we must, in the first place, premise,* as before stated, that we do not consider the complete conversion of organic nitrogen into ammonia by our method as being essential to its applicability for determining the relative quality of water, and that we rely simply upon the constancy of the ratio between the amount of albuminoid substance in the water and the quantity of ammonia produced.

The next point which we have to consider is the degree of accuracy attainable in estimating the carbon and nitrogen in the water-residue according to Frankland and Armstrong's method. In order to enable a judgment to be formed on this point, the authors gave ten examples, in which known weights of known substances were dissolved in distilled water with some carbonate of soda, or carbonate of lime, and the residues obtained, after treatment with SO_2 and evaporation, were burnt with chromate of lead. In the following tables we give the amounts of carbon taken, and the amounts ob-

* *Chem. Soc. Journal,* December 1867.

tained, also the amounts of nitrogen taken and obtained, and the errors applying to each experiment :—

Substance taken. Milligrm.		Carbon taken. Milligrm.	Error. Milligrm.	Carbon obtained. Milligrm.
I....35·2	Sugar	14·82	— 0·19	14·63
II....34·7	,,	14·60	— 0·74	13·86
III....11·4	,,	4·80	— 0·40	4·40
IV....12·2	,,	5·14	+ 0·16	5·30
V....11·5	,,	4·84	— 0·50	4·34
VI....10·	Urea	2·00	— 0·23	1·77
VII....10·25	,,	2·05	+ 0·06	2·11
VIII....10·4	,,	2·08	+ 0·31	2·39
IX....20·2	,,	4·04	+ 0·48	4·52
X....25·	Hippuric acid	15·08	— 1·22	13·86

	Nitrogen taken. Milligramme.	Error.	Nitrogen obtained. Milligramme.
V.............	2·46	+ 0·08	2·54
VI.............	4·66	— 0·03	4·63
VII.............	4·78	— 1·21	3·57
VIII.............	4·84	— 0·16	4·68
IX.............	9·42	— 0·55	8·87
X.............	1·95	+ 0·08	2·03

From these tables it will be seen that there is a deficit of carbon in six out of the ten determinations, and an excess in four of them. The greatest error in deficit is 1·22 milligrm.; the least error in deficit is 0·19 milligrms.; the mean error on the six determinations being 0·49 milligrms. In ordinary organic analyses, wherein 200 or 300 miligrms. of a substance such as sugar is taken, it is possible to obtain results which are accurate to within about one-tenth per cent. equal to an absolute error of 0·2 milligrm. of carbon. When smaller quantities of substance are analysed in the ordinary way, it is admitted that the degree of *per-centage accuracy* attainable is inferior.

The quantities of organic substance to which the results given by Frankland and Armstrong refer are from 10 to 35 milligrms., being about one-tenth as much as would be operated upon in an ordinary organic analysis. But the absolute error, as shown in the tables above, is from 0·2 to 1·2 milligrm. of carbon, so that while operating on smaller quantities there is no corresponding reduction of *absolute* error, and consequently Frankland and Armstrong's method, as exhibited by their own experiments, does not attain to a higher degree of accuracy than would be reached by ordinary organic analysis applied to very small quantities of organic substance.

The importance of this circumstance as regards the estimation of organic substance in a water will be appreciated when it is considered that the quantity of organic substance in a litre of water is seldom anything like so much as the quantities of sugar, &c., operated upon in the experiments given as indicative of the degree of accuracy attainable by Frankland and Armstrong's method.

From the results obtained for nitrogen it will be seen that out of the six experiments the results of four are in deficit, and those of two are in excess. These results apply to quantities of nitrogenous substance fully tenfold as large as those likely to be present in a litre of ordinary water. It therefore appears to us that, taking these data as representing the extent to which this method can be depended on for the determination of the minute quantities of carbon and nitrogen in a water-residue, it does not estimate quantities of nitrogen which fall short of half a milligramme. Now, on turning to the table of analyses at the end of the memoir,* it will be seen that the quantity of organic nitrogen per litre (and a litre is the quantity of water upon which the determinations were made) is represented as ranging from 0·00 through all varieties of intermediate value to 0·56 milligrm. These quantities of nitrogen, are, however, within the limits of error indicated by the experiments above referred to ; consequently, we cannot regard these results as representing differences of quality in the different kinds of water.

Having thus considered what are the capabilities of the method proposed by Frankland and Armstrong, we will now proceed to discuss the comparison which they institute between the results furnished by our method and those obtained by their own. In the first place, it will be seen that the differences between the results of the two

* *Chem. Soc. Journal*, March 1868, p. 108.

methods observed by Frankland and Armstrong, range from $+$ 0·05 to $-$ 0·52 of a milligramme, amounts which, as we have already shown, lie within the limits of experimental error.

From these differences, therefore, no conclusion of any kind can be drawn, and we consider it to be sufficiently evident that Frankland and Armstrong's method is incapable of testing the accuracy of the results obtained by our method as stated below.

	Milligramme per Litre.	
	Albuminoid NH₃. Wanklyn, Chapman, and Smith.	Organic Nitrogen. Frankland and Armstrong.
Bala Lake water, . . .	0·25	0·01
Loch Katrine water, . .	0·13	0·08
Manchester water, . . .	0·07	0·26
Thames water as delivered in London by the different companies at different dates, . .	0·06	0·48
	0·15	—
	0·12	—
	0·14	—
	0·20	—
New River water, . . .	0·084	0·14
	0·09	—
East London Water Company,	0·09	0·24
Caterham water, . . .	0·00	0·07

On general grounds, we are disposed to consider that the circumstance of Frankland and Armstrong's method being applicable only to the residue obtained by evaporation of a water, is a disadvantage, both as regards the time requisite for making an experiment, and on account of the probability of loss of

organic substance. But, in our opinion, the preliminary treatment of the water with SO_2 in order
to eliminate nitrogen existing as nitrates and
nitrites, comprises a source of error of a far more
serious character. There can be no question as to
the complete expulsion of CO_2 by this treatment,
and we therefore pass over that part of the subject.
But, as the nitrogen existing as nitrates in some
kinds of water is often much more than ten times
as much as the nitrogen existing in organic states
of combination, it will be manifest that the estimation of organic nitrogen by Frankland and Armstrong's method would become illusory, if only a
small portion of the nitrates were to escape decomposition.

On referring to Frankland and Armstrong's paper
it will be seen that the process for the destruction
of the nitrates and nitrites is as follows :—" 2 litres
are poured into a convenient stoppered bottle, and
60 c. c. of a recently prepared saturated solution of
sulphurous acid are added." . . . . . " One half of
this sulphurised water is now boiled for two or three
minutes, and unless it contained a considerable
amount of carbonates, 0·2 grm. of sodic sulphite is
to be added during the boiling, so as to secure the
saturation of the sulphuric acid formed during the
subsequent evaporation." The addition of "a couple
of drops" of solution of ferrous or ferric chloride is

also recommended, and the water is subsequently to be evaporated to dryness upon a steam or water-bath.

The authors remark further on, that " such an expulsion of the nitrogen of nitrates and nitrites is a remarkable reaction, and could scarcely have been predicted; indeed, it takes place to a very partial extent only when a nitrate is dissolved in water, and evaporated with excess of sulphurous acid in imitation of a natural water ; neither is the result very different when sodic chloride, or calcic or magnesic carbonate is added."

We agree with the authors in looking upon a total decomposition of a *nitrate* by a few minutes' boiling with a solution of sulphurous acid as remarkable. On the other hand, a decomposition of free *nitric acid* by sulphurous acid is what we should be quite prepared to expect.

When it is considered that a litre of many waters contains in solution sufficient free oxygen to generate 0·06 grm. of sulphuric acid by oxidation of the sulphurous acid, it will become obvious that, notwithstanding the addition of the 0·2 grm. of sodic sulphite, which is recommended in the case of those waters containing no considerable amount of carbonates, there will always be great danger of the water becoming strongly acid. The probability of the 30 c. c. of saturated solution of sulphurous acid containing sulphuric acid is also great; there is, moreover, the risk of absorption of oxygen, and consequent formation of sulphuric acid during the standing in the bottle, and during the boiling in the flask.

It is worthy of note, that the addition of the "couple of drops" of solution of ferric chloride, which the authors find to be so efficacious in rendering the decomposition of the nitrate complete, is equivalent to an addition of so much free acid.

Of the six experiments given by Frankland and Armstrong to illustrate the *complete* decomposition of the nitrates, (pp. 96 and 97,) the first one, the third and fourth, are instances in which, from the absence of any alkaline or earthy sulphite to take up the sulphuric acid, there must necessarily have been free nitric acid from the very beginning of the reaction.   In the second of these experiments, 10 c. c. of a solution of sodic sulphite (strength unknown) were added.   In the fifth experiment a natural water was taken, but no mention is made of the amount of carbonate of lime in it.   Only the sixth admits of discussion as a possible example of a complete reduction of nitrates without the charging of the water with free sulphuric acid.   In this experiment, ·01 grm. magnesia, 0·1 grm. calcic carbonate, 0·1 grm. sodic chloride, 0·01 grm. potassic chloride, 1 drop of solution of ferric chloride, 2 drops of solution of hydric sodic phosphate, 0·1 grm. potassic nitrate, and 15 c. c. of sulphurous acid solution were taken, and complete destruction of the nitrate was the result.

By calculation, it will be seen that the 10 milligrammes of magnesia and 100 milligrammes of carbonate of lime are equivalent to 122·5 milligrammes of sulphuric acid. Now the oxygen dissolved in the water cannot have been less than would suffice to form about 60 milligrammes of sulphuric acid, whilst 97 milligrammes of sulphuric acid would be set free by the reduction of the 100 milligrammes of nitrate of potash. Thus we should have about 34·5 milligrammes of free sulphuric acid as the final result of the reaction.

It would, therefore, appear that in Frankland and Armstrong's test-experiments, in which there was complete reduction of the nitrates, the circumstances were such as to give rise to free sulphuric acid as a final product.

We have made experiments in which care was taken to avoid the production of this acid as an ultimate product of the reaction, and have never succeeded, under such circumstances, in effecting a *complete* destruction of the nitrates. In some instances, iron and phosphates were present in the *natural* waters experimented upon; but still the destruction of the nitrates was incomplete.

The following experiments may be cited:—Water from a pump in Great Portland Street: 1 litre taken, 30 c. c. of a saturated solution of sulphurous acid was added, and then boiled for two minutes; evaporated on the steam-bath (one or two c. c. of a saturated solution of sulphurous acid being added four times during the evaporation.) Result: 14 milligrammes of HNO_3 were left in the residue. It is to be observed that this water contains both iron and phosphates. The water from a pump in Bartholomew Lane gave a similar result.

To another well-water a quantity of sulphite of protoxide of iron was added before submitting it to the action of sulphurous acid, and yet the residue contained much unreduced nitrates.

In another instance, 100 milligrammes HNO_3 (in the state of nitrate of potash,) 300 milligrammes Ca_2OCO_2, 100 milligrammes KCl, and about 400 milligrammes of phosphate of lime were put into half a litre of distilled water, boiled with 30 c. c. saturated solution of sulphurous acid, and evaporated to dryness. Result: 55 milligrammes of HNO_3 left undecomposed in the residue.

On repeating this experiment and substituting sulphite of protoxide of iron for the phosphate of lime, 69·5 milligrammes of HNO_3 was left undecomposed.

In another experiment a half-litre of New River

water was taken, saturated with carbonic acid, and then boiled, cooled in an atmosphere of that gas, and again boiled and cooled in carbonic acid; in this way it was insured that the water should be free from dissolved oxygen. 15 c. c. of a saturated solution of sulphurous acid, free from sulphuric acid, was then added, and the whole mixture boiled for two and a-half minutes and evaporated to dryness. The residue was found to contain much nitric acid. (These determinations of nitric acid were made by a modification of Schulze's aluminium-process, a description of which has been laid before the Chemical Society.)

From all these experiments two facts are very apparent. First, the operation of destroying the nitrates in water by means of sulphurous acid is a very uncertain one. Secondly, the operation, as practised by Frankland and Armstrong, gives rise to free sulphuric acid in the residue. We need hardly add that few organic substances will bear being heated to 100° C., with their own weight of sulphuric acid, without undergoing great decomposition.

We must here call attention to the circumstance that in the experiments made to determine the degree of accuracy attainable by Frankland and Armstrong's method of estimating carbon and nitrogen, no nitrates were added to the water, and as there was in all cases sufficient alkaline or calcareous sulphite to take up any sulphuric acid produced by oxidation of sulphurous acid by the oxygen dissolved in the water, there was no charring of the residue by sulphuric acid. In this very material condition, therefore, the trial experiments differ from those made to prove complete destruction of nitrates, and from operations on natural water.

In the case of water containing large quantities of organic nitrogen, as for example actual sewage, in which the amount of nitrogen would be capable of estimation by Frankland and Armstrong's method, we encounter another difficulty, due to the presence of ammonia in the water-residue.

In order to arrive at the organic nitrogen, it would then be necessary to make a determination of the free ammonia in the water, and to deduct the nitrogen corresponding to it from the total nitrogen of the residue. Owing, however, to the circumstance that ammonia would be lost by diffusion during the evaporation, (even in presence of an acid,) the water-residue will contain only a part of the original ammonia, and consequently an error would arise in deducting the amount of nitrogen corresponding to the original ammonia, from the total nitrogen of the residue.

This source of error would, of course, affect to some extent all determinations of organic nitrogen in waters containing free ammonia, but it would become important in the case of such waters as London well water, which often contains a considerable quantity of free ammonia.

We will conclude by giving in a tabular form a number of analyses made by our ammonia-method, showing the extreme constancy of the results. Each of the first five sets of analyses bracketed together was made on the same day, on the same sample of water.

Name of Water.	Quantity operated upon. Litres.	Milligrammes of Albuminoid Ammonia per litre of the water.
West Middlesex water.. {	$\frac{1}{2}$ 1 $1\frac{1}{2}$	0·070 0·065 0·07
Southwark and Vaux-hall .... {	1 1 $\frac{1}{2}$ $\frac{1}{4}$ $1\frac{1}{2}$	0·20 0·205 0·19 0·18 0·21
Well at Wimbledon...... {	1 $\frac{1}{2}$	0·15 0·16
Bishopsgate St. pump... (Mixed with equal volume) of distilled water ...... (	1 $\frac{1}{2}$ $\frac{1}{2}$ (= 1 litre of mixture.)	0·24 0·255 } 0·26
Thames water above { Hampton	$\frac{1}{2}$ 1 $1\frac{1}{2}$	0·21 0·21 0·205
Manchester water all (taken at same *date*, but) from different parts (of the town............. (	$\frac{1}{2}$ $\frac{1}{2}$ $\frac{1}{2}$	0·06 0·07 0·07
Edinburgh water taken (from a tap at the Uni-) versity on 18th and) 19th Sept. 1867......... (	1 $1\frac{1}{2}$ 1	0·075 0·063 0·070

London Institution.

INDEX.

_____

www.ingramcontent.com/pod-product-compliance
Lightning Source LLC
Chambersburg PA
CBHW021824190326
41518CB00007B/734